Study Guide for

Structure and Function
of the Human Body

Ruth L. Memmler, MD

Professor Emeritus
Life Sciences
formerly Coordinator
Health, Life Sciences, Nursing
East Los Angeles College
Los Angeles, California

•

Barbara Janson Cohen, BA, MSEd.

Assistant Professor
Delaware County Community College
Media, Pennsylvania

•

Dena Lin Wood, RN, MS

Staff Nurse
Verdugo Hills Visiting Nurse Association
Glendale, California

**Illustrated by Janice A. Schwegler, MS, CMI
and
Anthony Ravielli**

Study Guide for
Structure and Function
of the Human Body

Sixth Edition

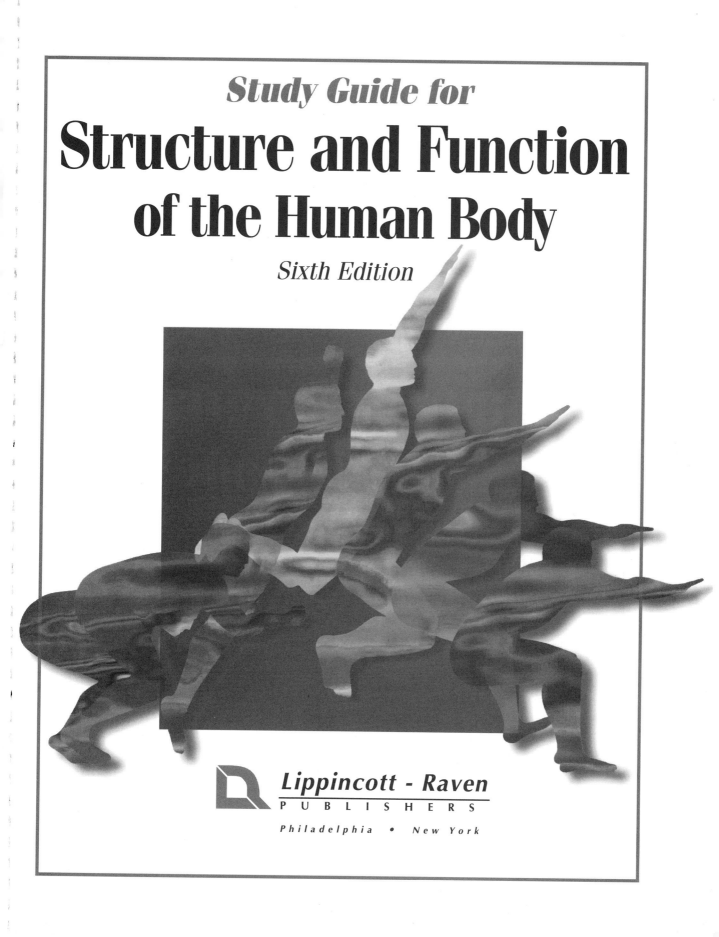

Lippincott - Raven
P U B L I S H E R S

Philadelphia • *New York*

Acquisitions Editor: Andrew Allen
Editorial Assistant: Laura Dover
Project Editor: Barbara Ryalls
Indexer: Alberta Morrison
Design Coordinator: Doug Smock
Cover Designer: Lawrence R. Didona
Production Manager: Helen Ewan
Production Coordinator: Nannette Winski

6th Edition

6   5   4   3   2   1

Any procedure or practice described in this book should be applied by the health care practi-
tioner under appropriate supervision in accordance with professional standards of care used
with regard to the unique circumstances that apply in each practice situation. Care has been
taken to confirm the accuracy of information presented and to describe generally accepted
practices. However, the authors, editors, and publisher cannot accept any responsibility for
errors or omissions or for any consequences from application of the information in this book
and make no warranty, express or implied, with respect to the contents of the book.

Every effort has been made to ensure drug selections and dosages are in accordance with
current recommendations and practice. Because of ongoing research, changes in government
regulations, and the constant flow of information on drug therapy, reactions, and interactions,
the reader is cautioned to check the package insert for each drug for indications, dosages,
warnings, and precautions, particularly if the drug is new or infrequently used.

Preface

Study Guide for Structure and Function of the Human Body, Sixth Edition, assists the beginning student to learn basic information required in the health occupations. Though it will be most effective when used in conjunction with the 6th edition of *Structure and Function of the Human Body*, the study guide is also applicable to other textbooks on basic anatomy and physiology.

The questions in this edition reflect changes in the text, and the labeling exercises include many of the new illustrations. The *Practical Applications* portions of the study guide use clinical situations to test the student's understanding of a subject.

The exercises are designed to help in student learning, not merely to test knowledge. A certain amount of repetition has been purposely incorporated as a means of reinforcement. Matching questions require the student to write out complete answers to give practice in spelling as well as recognition of terms. New in this edition are true–false questions, for which the false statements must be corrected. Short essay questions are also given for each chapter. The essay answers provided are examples of suitable responses, but other presentations of the material are acceptable. Multiple-choice questions are included to give the student practice in taking standardized tests. All answers to the *Study Guide* questions are in the *Instructor's Manual* that accompanies the text.

Contents

UNIT **I**

The Body as a Whole 1

1 Introduction to the Human Body   3
2 Chemistry, Matter, and Life   17
3 Cells and Their Functions   27
4 Tissues, Glands, and Membranes   37
5 The Skin   51

UNIT **II**

Movement and Support 59

6 The Skeleton: Bones and Joints   61
7 The Muscular System   85

UNIT **III**

Coordination and Control 97

8 The Nervous System: The Spinal Cord and Spinal Nerves   99
9 The Nervous System: The Brain and Cranial Nerves   113
10 The Sensory System   127
11 The Endocrine System: Glands and Hormones   141

UNIT IV

Circulation and Body Defense 151

12 The Blood **153**

13 The Heart **163**

14 Blood Vessels and Blood Circulation **177**

15 The Lymphatic System and Immunity **197**

UNIT V

Energy: Supply and Use 209

16 Respiration **211**

17 Digestion **225**

18 Metabolism, Nutrition, and Body Temperature **243**

19 The Urinary System and Body Fluids **251**

UNIT VI

Perpetuating Life 269

20 The Male and Female Reproductive Systems **271**

21 Development and Heredity **287**

22 Biologic Terminology **301**

Study Guide for

Structure and Function
of the Human Body

The Body as a Whole

1. INTRODUCTION TO THE HUMAN BODY
2. CHEMISTRY, MATTER, AND LIFE
3. CELLS AND THEIR FUNCTIONS
4. TISSUES, GLANDS, AND MEMBRANES
5. THE SKIN

Memmler, RL, Cohen, BJ, Wood, DL. *STUDY GUIDE FOR STRUCTURE AND FUNCTION OF THE HUMAN BODY*, 6/e, © 1996, Lippincott-Raven Publishers

CHAPTER 1

Introduction to the Human Body

I. Overview

Living things are organized from simple to complex levels. The simplest living form is the *cell,* the basic unit of life. Specialized cells are grouped into *tissues* that in turn are combined to form *organs;* these organs form *systems.*

The systems include the skeletal system, the framework of the body; the muscular system, which moves the bones; the circulatory system, consisting of the heart and blood vessels that transport vital substances; the digestive system, which converts raw food materials into products usable by cells; the respiratory system, which adds oxygen to the blood and removes carbon dioxide; the integumentary system, the body's covering; the urinary system, which removes wastes and excess water; the nervous system, the central control system that includes the organs of special sense; the endocrine system, which produces the regulatory hormones; and the reproductive system, by which new individuals of the species are produced.

All the cellular reactions that sustain life together make up *metabolism,* which can be divided into *catabolism* and *anabolism.* In catabolism, complex substances, such as the nutrients from food, are broken down into smaller molecules with the release of energy. This energy is stored in the compound *ATP* (adenosine triphosphate) for use by the cells. In anabolism, simple compounds are built into substances needed for cell activities.

All the systems work together to maintain a state of balance or *homeostasis.* The main mechanism for maintaining homeostasis is negative feedback, by which the state of the body acts to keep conditions within set limits.

It is essential that a special set of directional terms be learned to locate parts and to relate the various parts to each other. Several *planes of division* represent different directions in which cuts can be made through the body. Separation of the body into areas and regions, together with the use of the special terminology for directions and locations, makes it possible to describe an area within the human body with great accuracy.

The large internal spaces of the body are the *cavities,* in which various organs are located. The *dorsal cavity* is subdivided into the cranial cavity and the spinal canal. The *ventral cavity* is subdivided into the thoracic and abdominopelvic cavities. Imaginary lines are used to divide the abdomen into regions for study and diagnosis.

The metric system is used for all scientific measurements. This system is easy to use because it is based on multiples of 10.

II. Topics for Review

A. Body systems
B. Body processes
C. Body directions
D. Body cavities
    1. Dorsal and ventral cavities
    2. Regions of the abdomen
E. The metric system

III. Matching Exercises

Matching only within each group, write the answers in the spaces provided.

Group A

organ	anatomy	physiology
system	tissue	cell

1. A specialized group of cells   _____

2. The study of how the body functions   _____

3. A combination of tissues that function together   _____

4. A group of organs functioning together for the same general purpose   _____

5. The basic unit of life   _____

6. The study of body structure   _____

Group B

epigastric	diaphragm	umbilicus
lateral	frontal	thoracic
transverse		

1. A plane that divides the body into anterior and posterior parts   _____

2. A directional term that means away from the midline (toward the side)   _____

3. A plane that divides the body into superior and inferior parts   _____

4. Term describing the central region of the abdomen just below the breast bone   _____

5. Another name for the navel   _____

6. A term that describes the uppermost (chest) portion of the ventral body cavity   _____

7. The muscular partition between the two main ventral body cavities   _____

Group C

caudal	cranial	posterior
sagittal	distal	transverse

1. A term that means away from the origin of a part   _____

2. A term that indicates a location toward the back   _____

3. A word that means nearer to the sacral (lowermost) region of the spinal cord   _____

4. A word that means nearer to the head   _____

5. A plane of division that is also described as a horizontal or cross section   _____

6. A plane that divides the body into left and right parts   _____

Group D

urinary system	integumentary system	skeletal system
endocrine system	reproductive system	respiratory system

1. The system that includes the hair, nails, and skin   _____

2. The system made up of the bones and joints   _____

3. Another name for the excretory system   _____

4. The system of scattered organs that produce hormones   _____

5. The system that includes the sex organs   _____

6. The system made up of the lungs and the passages leading to and from the lungs   _____

IV. Multiple Choice

Select the best answer and write the letter of your choice in the blank.

1. Another name for a frontal plane is

    a. distal
    b. horizontal
    c. transverse
    d. coronal
    e. sagittal

1. _____

2. The term *ventral* means

    a. toward the belly surface
    b. posterior
    c. in the dorsal body cavity
    d. farther from the origin of a structure
    e. nearer to the back

2. _____

3. Anabolism produces

    a. simple compounds from more complex compounds
    b. carbon dioxide
    c. energy
    d. digested foods
    e. complex materials needed for body functions

3. _____

4. Fluids located outside the cells are described as

    a. lateral
    b. intracellular
    c. superior
    d. extracellular
    e. frontal

4. _____

5. A system that controls and coordinates the body is the

    a. circulatory system
    b. urinary system
    c. nervous system
    d. digestive system
    e. skeletal system

5. _____

6. The cavity below the abdominal cavity is the

    a. dorsal cavity
    b. pelvic cavity
    c. thoracic cavity
    d. cranial cavity
    e. frontal cavity

6. _____

7. A plane that divides the body into upper and lower parts is called a

    a. transverse plane
    b. frontal plane

7. _____

   c. midsagittal plane
   d. superior plane
   e. sagittal plane

V. Labeling

For each of the following illustrations, write the name or names of each labeled part on the numbered lines.

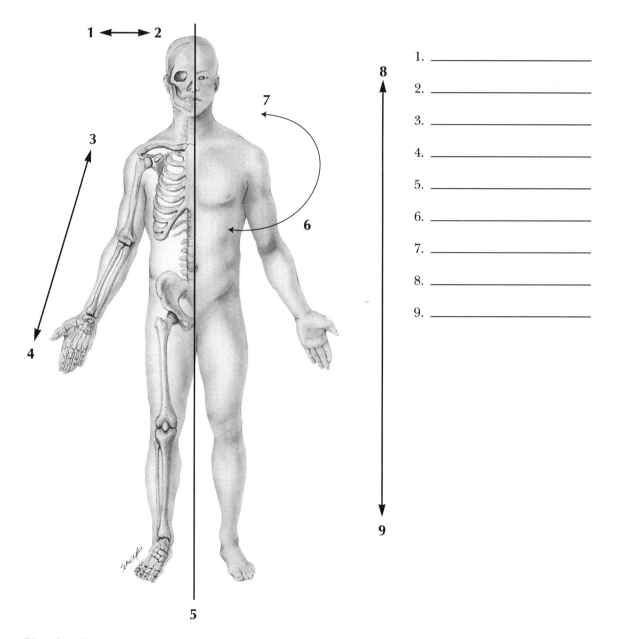

1. _____

2. _____

3. _____

4. _____

5. _____

6. _____

7. _____

8. _____

9. _____

Directional terms

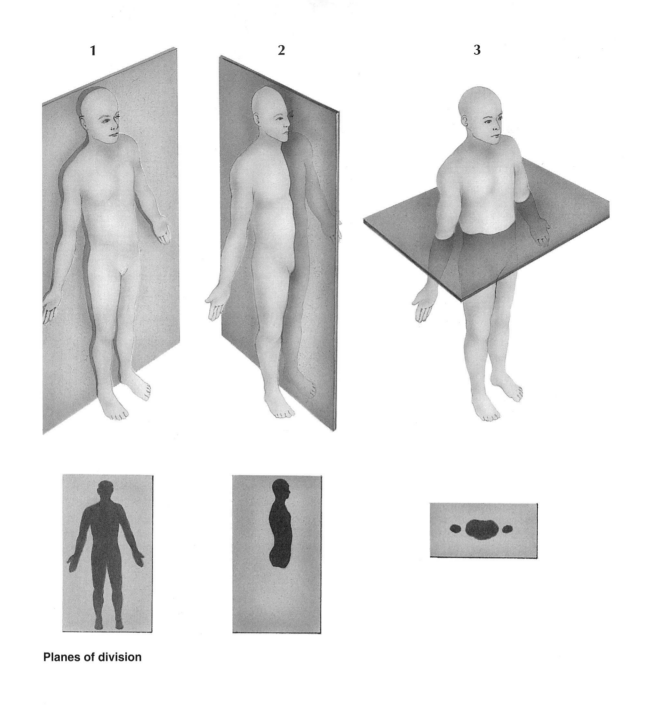

Planes of division

1. _____   3. _____

2. _____

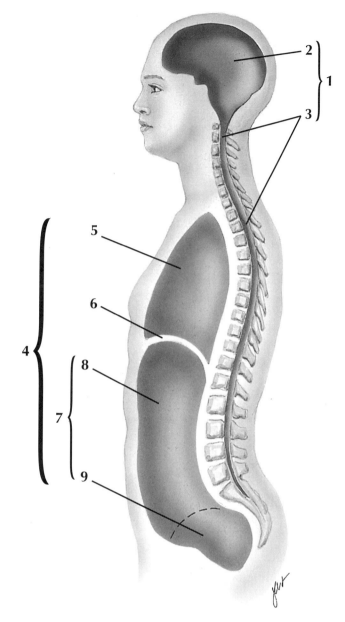

Side view of body cavities

1. _____   6. _____

2. _____   7. _____

3. _____   8. _____

4. _____   9. _____

5. _____

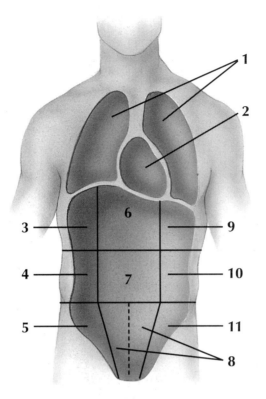

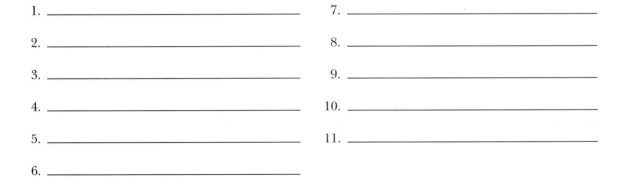

Front view of thoracic cavity and nine regions of the abdomen

1. _____   7. _____

2. _____   8. _____

3. _____   9. _____

4. _____   10. _____

5. _____   11. _____

6. _____

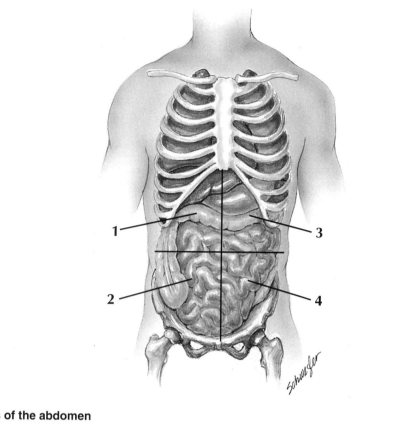

Quadrants of the abdomen

1. _____ 3. _____

2. _____ 4. _____

VI. True-False

For each question, write T for true and F for false in the blank to the left of each number. If a statement is false, correct it by replacing the <u>underlined</u> term and write the correct statement in the blanks below the question.

_____ 1. The word <u>medial</u> indicates nearness to the midsagittal plane.

_____

_____

_____ 2. The term <u>proximal</u> means away from the point of origin.

_____

_____

_____ 3. The term <u>caudal</u> means toward the head.

_____

_____

_____ 4. Another term for dorsal is <u>anterior</u>.

_____

_____

_____ 5. In <u>catabolism</u>, nutrients are broken down into simpler compounds.

_____

_____

VII. Completion Exercise

Group A

Write the word or phrase that correctly completes each sentence.

1. All the chemical reactions that sustain life make up
   _____

2. The main energy compound of the cell is
   _____

3. Negative feedback is a mechanism for maintaining an
   internal state of balance known as
   _____

4. A specialized group of cells make up a(n)
   _____

5. Regions and directions in the body are described according to
   the position in which the body is upright, with the palms
   facing forward. This is called the
   _____

6. The midline plane that divides the body into right and
   left halves is the
   _____

7. The plane that divides the body into anterior and posterior
   parts is the
   _____

8. The space that encloses the brain and spinal cord forms a
   continuous cavity called the
   _____

9. The space that houses the brain is the
   _____

10. The elongated canal that contains the spinal cord is known
    as the
    _____

11. The ventral body cavity includes an upper space containing the lungs, the heart, and the large blood vessels, which is called the

_____

12. The ventral body cavity that contains the stomach, most of the intestine, the liver, and the spleen is the

_____

13. The main large ventral body cavities are separated from each other by a muscular partition called the

_____

14. The abdomen may be subdivided into nine regions, including three along the midline. The uppermost of these midline areas is the

_____

15. The abdomen may be divided into four regions, each of which is called a(n)

_____

Group B

Write the word that correctly completes each sentence about the metric system.

1. The standard unit for measurement of volume, slightly greater than a quart, is a(n)

_____

2. The number of grams in a kilogram is

_____

3. The number of centimeters in an inch is

_____

4. The number of milliliters in 0.5 liters is

_____

5. The standard unit for measurement of length is the

_____

VIII. Practical Applications

Study each discussion. Then write the appropriate word or phrase in the space provided.

Group A

1. The gallbladder is located just below the liver. The directional term that describes the position of the gallbladder with regard to the liver is

_____

2. The kidneys are located behind the other abdominal organs. This position may be described as

_____

3. The nails are located farthest from the origin of the fingers and toes. This position is described as

_____

4. The entrance to the stomach is nearest the point of origin or beginning of the stomach, so this part is said to be

_____

5. The ears are located away from the midsagittal plane or toward the side, so they are described as being

_____

6. The head of the pancreas is nearer the midsagittal plane than its tail portion, so the head part is more

_____

7. The diaphragm is above the abdominal organs; it may be described as

_____

8. If the abdomen is divided into four regions, the left ovary is in the _____

Group B

Answer the questions based on the nine divisions of the abdomen.

1. Mr. A had an appendectomy. The appendix is in the lower right side of the abdomen. In which of the nine abdominal regions is it located?

_____

2. Mrs. D had a history of gallstones. The operation to remove these stones involved the upper right part of the abdominal cavity. Which of the nine abdominal regions was this?

_____

3. Miss C was injured in an automobile accident. In addition to a number of fractures, she suffered a ruptured urinary bladder. The region involved, in the lower midline part of the abdomen, was the

_____

4. Mr. B required an extensive exploratory operation that necessitated incision through the navel. This portion of the abdomen is in what region?

_____

IX. Short Essays

1. Define metabolism and describe the two phases of metabolism.

_____

_____

_____

_____

2. Explain homeostasis and how negative feedback works to maintain homeostasis.

_____

_____

_____

_____

3. Explain why specialized terms are needed to indicate different positions and directions within the body.

_____

_____

_____

_____

Memmler, RL, Cohen, BJ, Wood, DL. *STUDY GUIDE FOR STRUCTURE AND FUNCTION OF THE HUMAN BODY*, 6/e, © 1996, Lippincott-Raven Publishers

CHAPTER 2

Chemistry, Matter, and Life

I. Overview

Chemistry is the physical science that deals with the composition of matter. To appreciate the importance of chemistry in the field of health, it is necessary to know about atoms, molecules, elements, compounds, and mixtures. Though exceedingly small particles, atoms possess a definite structure: the *nucleus* contains *protons* and *neutrons,* and surrounding the nucleus are the *electrons.* An *element* is a substance consisting of just one type of atom. Isotopes are forms of an element that differ in atomic weight. Those isotopes that give off radiation are said to be *radioactive.* Union of two or more atoms produces a *molecule;* the atoms may be alike (such as the oxygen molecule) or different (sodium chloride, for example), and in the latter case the substance is called a *compound.* To go a step further, a combination of compounds, each of which retains its separate properties, is a *mixture* (salt water is one example). Chemical compounds are constantly being formed, altered, broken down, and recombined into other substances.

Water is a vital substance composed of hydrogen and oxygen. It makes up more than half of the body and is needed as a solvent and a transport medium. Hydrogen, oxygen, carbon, and nitrogen are the elements that constitute about 99% of living matter, whereas calcium, sodium, potassium, phosphorus, sulfur, chlorine, and magnesium account for most of the remaining 1%. Proteins, carbohydrates, and lipids are among the compounds formed from these elements. An important group of proteins is the *enzymes,* which function as catalysts in metabolism.

II. Topics for Review

A. Atoms and elements
B. Molecules and compounds
C. Water, solutions, and suspensions
D. Chemical bonds
E. Acids, bases, and buffers
F. Organic compounds

III. Matching Exercises

Matching only within each group, write the answers in the spaces provided.

Group A

element   organic   nucleus
chemistry   atoms   isotopes

1. The smallest complete units of matter   _____

2. The science that deals with the composition of all matter   _____

3. Elements existing in forms that are alike in their chemical
    reactions but that differ in weight   _____

4. A substance composed of one type of atom   _____

5. The part of the atom containing most of its mass including
    protons and neutrons   _____

6. Term for the chemical compounds that characterize
    living things   _____

Group B

radioactivity   compounds   mixture
molecule   electrons   protons
neutrons

1. The positively charged particles inside the atomic nucleus   _____

2. The noncharged particles within the atomic nucleus   _____

3. The negatively charged electric particles outside the
    atomic nucleus   _____

4. The word that refers to emission (giving off) of rays from
    disintegrating isotopes   _____

5. The unit formed by the union of two or more atoms   _____

6. Substances that result from the union of two or more
    different atoms   _____

7. A combination of compounds, each of which remains intact and retains its properties

_____

Group C

carbohydrates   cations   acid
electrolytes   anions   water
proteins   buffer   pH

1. Positively charged ions

_____

2. Negatively charged ions

_____

3. Compounds that form ions when in solution

_____

4. Organic compounds that contain nitrogen in addition to carbon, oxygen, and hydrogen

_____

5. The category of organic compounds that includes simple sugars and starches

_____

6. A substance that helps to maintain a stable hydrogen ion concentration in a solution

_____

7. The universal solvent

_____

8. The symbol for hydrogen ion concentration

_____

9. A substance that donates a hydrogen ion to another substance

_____

Group D

elements   covalent   phospholipids
suspension   colloidal   amino acid
atomic number   solution

1. Nitrogen, carbon, hydrogen, and oxygen, for example

_____

2. It identifies each element

_____

3. A building block of proteins

_____

4. The group of lipids that contains phosphorus in addition to carbon, hydrogen, and oxygen

_____

5. Term for a chemical bond formed by the sharing of electrons

_____

6. A mixture in which substances will settle out unless the mixture is shaken

_____

7. Cytoplasm and blood plasma are examples of this type of suspension

_____

8. A mixture in which the components remain evenly distributed, such as salt water _____

Group E

ionic   carbon   neutral
neutrons   lipid   enzyme

1. Term that describes a pH of 7.0 _____

2. A type of protein that acts as a catalyst in metabolic reactions _____

3. Another name for a fat _____

4. The type of bond that forms an electrolyte _____

5. Particles that vary in different isotopes _____

6. The element that is the basis of organic chemistry _____

IV. Multiple Choice

Select the best answer and write the letter of your choice in the blank.

1. Which of the following is <u>not</u> a common chemical element in the body?   1. _____

   a. nitrogen
   b. carbon
   c. glucose
   d. oxygen
   e. hydrogen

2. Which of the following statements is <u>not</u> true of water?   2. _____

   a. It contains hydrogen and oxygen.
   b. It is a compound.
   c. It is stable at ordinary temperatures.
   d. It is organic.
   e. It can dissolve many different substances.

3. A mixture that does not separate because the particles in the mixture are so small is a(n)   3. _____

   a. colloidal suspension
   b. solvent
   c. radioactive isotope
   d. true solution
   e. enzyme

4. Which of the following statements is <u>not</u> true of electrolytes?   4. _____

   a. They conduct an electrical current in solution.
   b. They separate into charged particles in solution.

c. They are compounds.

d. They are found in body fluids.

e. They are insoluble in water

5. The pH scale ranges from

5. _____

a. 5–10   d. 3–6

b. 0–14   e. 0–20

c. 10–20

6. Which of the following statements is <u>not</u> true of enzymes?

6. _____

a. Their names usually end in the suffix *-ase*.

b. Their shape is important in reactions.

c. They are not changed in reactions.

d. They are not damaged by extreme temperatures.

e. They speed up the rate of a reaction.

V. Labeling

For each of the following illustrations, write the name or names of each labeled part on the numbered lines.

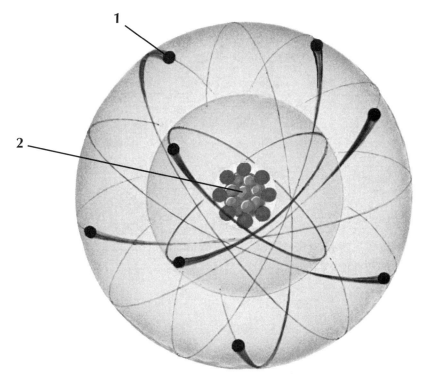

Oxygen atom

1. _____   2. _____

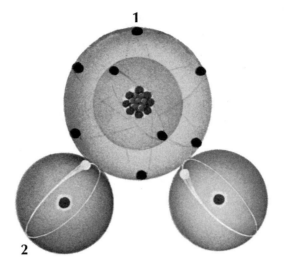

Molecule of water

1. _____   2. _____

VI. True-False

For each question, write T for true and F for false in the blank to the left of each number. If a statement is false, correct it by replacing the <u>underlined</u> term and write the correct statement in the blanks below the question.

_____ 1. The noncharged particles in the nucleus of an atom are called <u>protons</u>.

_____

_____

_____ 2. The first energy level of an atom can hold <u>eight</u> electrons.

_____

_____

_____ 3. An atom that normally loses electrons to attain a complete outer energy level is called a <u>metal</u>.

_____

_____

_____ 4. The smallest unit of a compound is an <u>atom</u>.

_____

_____

_____ 5. A positively charged ion is an <u>anion</u>.

_____

_____

_____ 6. A <u>covalent</u> bond is one formed by the sharing of electrons.

_____

_____

_____ 7. A pH of <u>7.0</u> is neutral.

_____

_____

_____ 8. A substance that can accept a hydrogen ion is an <u>acid</u>.

_____

_____

_____ 9. Amino acids are the building blocks of <u>proteins</u>.

_____

_____

VII. Completion Exercise

Write the word or phrase that correctly completes each sentence.

1. The four elements that make up about 99% of living cells are oxygen, carbon, and hydrogen plus _____

2. Many people keep a shaker of salt on the table. Salt is an example of a combination of two different elements. Such a combination is called a(n) _____

3. Compounds first found in living organisms, as for example starch in potatoes, are classified as _____

4. An element that is part of the air we breathe also is part of the water we drink. This element is _____

5. The smallest particle of a compound that would still have the properties of that compound is a(n) _____

6. The salt in salt water will regain its properties if the water is boiled away. Since water and salt do not combine chemically, this solution is an example of a(n) _____

7. If an electron is added to or removed from an atom it becomes a charged particle, also known as a(n) _____

8. Numerous essential body activities are possible owing to the property of certain compounds to form ions when in solution. Such compounds are called _____

9. The name given to a chemical system that prevents changes in hydrogen ion concentration is _____

10. Metabolic reactions require organic catalysts that are called _____

VIII. Practical Applications

Study each discussion. Then write the appropriate word or phrase in the space provided. The following medical tests are based on principles of chemistry and physics.

1. Mr. B complained of shortness of breath. Several studies were done including a visible tracing of the electric currents produced by his heart muscle. Such a record is called a(n) _____

2. Joan, age 4, was brought to the clinic by her mother because she experienced attacks of fainting and unconsciousness. As an aid in diagnosis, a graphic record of her brain's electric current was obtained. This brain wave record is called a(n) _____

3. A routine test done on Ms. J showed glucose in her urine—an abnormal finding. Glucose is one of a group of organic compounds classified as _____

4. Mr. K's urinalysis showed the presence of albumin. Albumin is an example of compounds found in the body that contain nitrogen, carbon, hydrogen, and oxygen. These compounds are classified as _____

5. Mrs. L, age 72, was brought to the clinic with symptoms of decreased function of many systems. Her history revealed poor fluid intake for several weeks. Her symptoms were due to a shortage of the most abundant compound in the body, which is _____

IX. Short Essays

1. Describe the structure of an atom.

   _____

   _____

   _____

   _____

2. What properties make water an ideal medium for living cells?

   _____

   _____

   _____

   _____

3. Why is carbon the basis of organic chemistry?

   _____

   _____

   _____

   _____

4. Why is the shape of an enzyme important in its function?

   _____

   _____

   _____

   _____

Memmler, RL, Cohen, BJ, Wood, DL. *STUDY GUIDE FOR STRUCTURE AND FUNCTION OF THE HUMAN BODY*, 6/e, © 1996, Lippincott-Raven Publishers

CHAPTER 3

Cells and Their Functions

I. Overview

The cell is the basic unit of life; all life activities result from the activities of cells. The study of cells began with the invention of the light microscope and has continued with the development of electron microscopes. Cell functions are carried out by specialized structures within the cell called *organelles.* These include the nucleus, ribosomes, mitochondria, Golgi apparatus, and endoplasmic reticulum (ER).

An important cell function is the manufacture of *proteins,* including enzymes (organic catalysts). Protein manufacture is carried out by the ribosomes in the cytoplasm according to information coded in the deoxyribonucleic acid (DNA) of the nucleus. DNA is also involved in the process of cell division or *mitosis.* Before cell division can occur, the DNA must double itself so that each daughter cell produced by mitosis will have exactly the same kind and amount of DNA as the parent cell.

The cell membrane is important in regulating what enters and leaves the cell. Some substances can pass through the membrane by *diffusion,* which is simply the movement of molecules from an area where they are in higher concentration to an area where they are in lower concentration. The diffusion of water through the cell membrane is termed *osmosis.* Because water can diffuse very easily across the membrane, cells must be kept in solutions that have the same concentrations as the cell fluid. If the cell is placed in a solution of higher concentration (a hypertonic solution), it will shrink; in a solution of lower concentration (a hypotonic solution), it will swell and may burst. The cell membrane can also selectively move substances into or out of the cell by *active transport,* a process that requires energy (ATP) and carriers. Large particles and droplets of fluid are taken in by the processes of *phagocytosis* and *pinocytosis.*

II. Topics for Review

A. Cell structure
B. Protein synthesis
C. Cell division (mitosis)
D. Movement of materials across the cell membrane

III. Matching Exercises

Matching only within each group, write the answers in the spaces provided.

Group A

| filtration | isotonic | diffusion |
| osmosis | mitosis | active transport |

1. The process of body cell division   _____

2. The spread of molecules throughout an area   _____

3. The process by which water molecules diffuse through the cell membrane   _____

4. The passage of solutions through a membrane as a result of mechanical force   _____

5. Term for a solution that has the same concentration of molecules as the fluids within the cell   _____

6. The process by which the cell uses energy to move substances across the membrane   _____

Group B

mitochondria	ribosomes	cilia
lysosomes	nucleolus	cell membrane
flagellum	endoplasmic reticulum	

1. A system of membranes throughout the cell   _____

2. A small globule within the nucleus   _____

3. Small bodies in the cytoplasm that act in the manufacture of proteins   _____

4. The outer covering of the cell   _____

5. The organelles that convert energy to ATP   _____

6. A long whiplike extension used in cell locomotion   _____

7. Small bodies in the cell that contain digestive enzymes   _____

8. Small hairlike projections from the cell used to create movement around the cell

_____

Group C

ATP   nucleotide   centriole
proteins   genes   RNA
DNA

1. The chemical in the nucleus that makes up the chromosomes

_____

2. The organelle that is active in cell division

_____

3. The main energy compound of the cell

_____

4. A building block of nucleic acids

_____

5. The hereditary factors in the cell

_____

6. The nucleic acid that carries information from the nucleus to the ribosomes

_____

Group D

osmotic pressure   pinocytosis   proteins
hypertonic   hypotonic   prophase

1. Term for a solution that is less concentrated than the fluid within the cell

_____

2. The process by which a cell takes in droplets

_____

3. The force that draws water into a solution

_____

4. Term for a solution with a salt concentration greater than 0.9%

_____

5. The first stage of mitosis

_____

6. The substance manufactured according to the DNA code

_____

IV. Multiple Choice

Select the best answer and write the letter of your choice in the blank.

1. Which of the following statements is <u>not</u> true of the cell membrane?   1. _____
   a. It is composed mainly of lipids and proteins.
   b. It regulates what enters and leaves the cell.
   c. It protects the cell.
   d. It is semipermeable.
   e. It keeps nutrients out.

2. Which of the following statements is <u>not</u> true of mitosis?   2. _____

    a. The original cell produces two identical daughter cells.
    b. It follows duplication of DNA in the nucleus.
    c. It occurs at the same rate in all cells.
    d. It involves the centrioles and a spindle.
    e. It results in equal division of the chromosomes.

3. The stage of mitosis during which the chromosomes line up across the
spindle is called   3. _____

    a. metaphase
    b. anaphase
    c. prophase
    d. telophase
    e. none of the above

4. The substance that moves most rapidly through the cell membrane is   4. _____

    a. sucrose
    b. glucose
    c. water
    d. lipid
    e. DNA

5. Which of the following are required for active transport?   5. _____

    a. vesicles and cilia
    b. carriers and ATP
    c. mitosis and diffusion
    d. osmotic pressure and centrioles
    e. osmosis and lysosomes

6. Which of the following solutions is isotonic for body cells?   6. _____

    a. 5% salt
    b. 0.9% salt
    c. 10% glucose
    d. 10% dextrose
    e. distilled water

V. Labeling

For each of the following illustrations, write the name or names of each labeled part on the numbered lines.

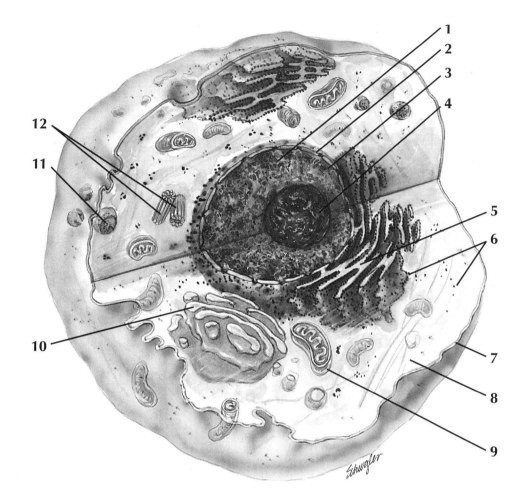

Diagram of a typical animal cell

1. _____	7. _____
2. _____	8. _____
3. _____	9. _____
4. _____	10. _____
5. _____	11. _____
6. _____	12. _____

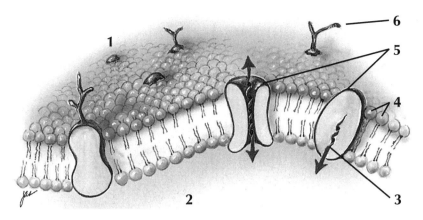

Structure of the cell membrane

1. _____ 4. _____

2. _____ 5. _____

3. _____ 6. _____

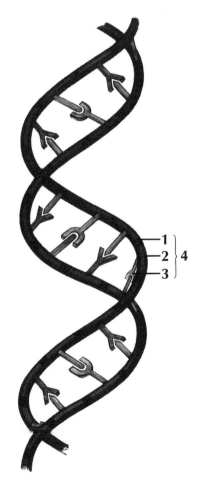

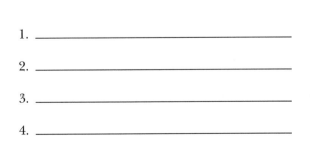

1. _____

2. _____

3. _____

4. _____

The basic structure of a DNA molecule

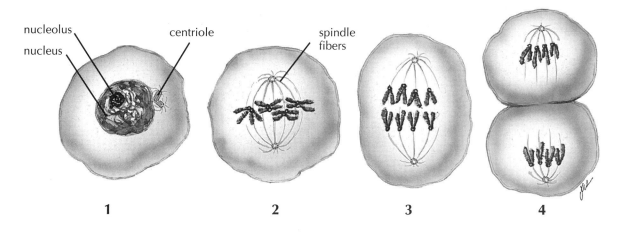

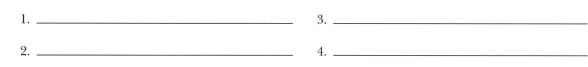

The stages of mitosis

1. _____   3. _____

2. _____   4. _____

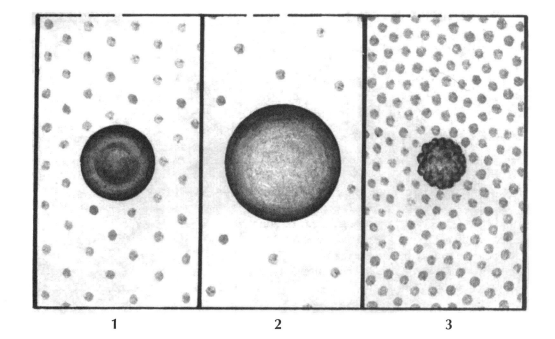

Osmosis

1. _____   3. _____

2. _____

VI. True-False

For each question, write T for true and F for false in the blank to the left of each number. If a statement is false, correct it by replacing the underlined term and write the correct statement in the blanks below the question.

_____ 1. A long, whiplike extension from the cell is called a <u>flagellum</u>.

_____

_____

_____ 2. The chromosomes are made of <u>RNA</u>.

_____

_____

_____ 3. A solution that is less concentrated than the intracellular fluid is described as <u>hypertonic</u>.

_____

_____

_____ 4. A 5% dextrose solution is <u>isotonic</u> for body cells.

_____

_____

_____ 5. A cell placed in a hypertonic solution will <u>shrink</u>.

_____

_____

_____ 6. In the <u>telophase</u> stage of mitosis, the duplicated chromosomes separate and move toward opposite ends of the cell.

_____

_____

VII. Completion Exercise

Write the word or phrase that correctly completes each sentence.

1. The metric unit that is used to measure cells is the _____

2. The substance that fills the cell and holds the cell contents is the _____

3. Small structures within a cell that perform special functions are called _____

4. The cell membrane uses energy to move materials from low concentration to a higher concentration (opposite to the direction they would normally flow by diffusion). The membrane is therefore described as _____

5. The control center of the cell, which contains the chromosomes, is the _____

6. Chromosomes are composed mainly of _____

7. The main energy compound of the cell is _____

8. Cells engulf large particles by the process of _____

9. The process of cell division is called _____

10. The period during which chromosomes duplicate occurs between mitoses and is called _____

11. The number of daughter cells formed when a cell undergoes mitosis is _____

VIII. Practical Applications

Study each discussion. Then write the appropriate word or phrase in the space provided. Observations you might make while touring a laboratory include the following:

1. The janitor in the laboratory was using a cleaning solution that contained ammonia. This activity would cause ammonia molecules to spread throughout the room. The movement of molecules from an area of high concentration to other areas where concentration is low is called _____

2. One of the laboratory technicians was trying to separate solid particles from a liquid mixture. He poured the mixture into a paper-lined funnel. The liquid flowed through the funnel while the solids remained behind on the paper. This process is called _____

3. A laboratory worker was carefully measuring certain salts to prepare a normal saline solution. Normal saline is used to replace lost body fluids because the concentration is the same as that inside the cells. Such a solution is said to be _____

4. While doing a complete blood count, a technician noted that some of the red blood cells had ruptured. The solutions used were tested to determine whether they were too dilute. When a red blood cell bursts because it is placed in a solution that is too dilute, it is said to _____

5. A student was learning how to do blood smears. On examination of the blood with the microscope, he found that many red blood cells appeared shrunken. The explanation was that he was proceeding so slowly that the liquid part of the blood was evaporating, leaving a more highly concentrated solution. Such a solution is described as being _____

6. The genetic material in a blood sample was being examined for abnormality of the chromosomes. For these tests, the chromosomes are studied at the time of cell division. The name for this process of cell division is _____

IX. Short Essays

1. Name the two types of nucleic acids and briefly describe how they act in the cell.

   _____

   _____

   _____

   _____

2. Explain why the cell membrane is described as selectively permeable.

   _____

   _____

   _____

3. Explain why it is important to keep cells in a solution that has the same concentration as the intracellular fluids.

   _____

   _____

   _____

Memmler, RL, Cohen, BJ, Wood, DL. *STUDY GUIDE FOR STRUCTURE AND FUNCTION OF THE HUMAN BODY*, 6/e, © 1996, Lippincott-Raven Publishers

CHAPTER 4

Tissues, Glands, and Membranes

I. Overview

The cell is the basic unit of life. Individual cells are grouped according to function into *tissues*. The four main groups of tissues include *epithelial tissue,* which forms glands, covers surfaces, and lines cavities; *connective tissue,* which gives structure and holds all parts of the body in place; *muscle tissue,* which produces movement; and *nervous tissue,* which conducts nerve impulses.

The simplest combination of tissues is a *membrane*. Membranes serve several purposes, a few of which are mentioned here: they may serve as dividing partitions, may line hollow organs and cavities, and may anchor various organs. Membranes that have epithelial cells on the surface are referred to as *epithelial membranes*. Two types of epithelial membranes are serous membranes, which line body cavities and cover the internal organs, and mucous membranes, which line passageways leading to the outside.

Glands produce substances for use by other cells and tissues. *Exocrine glands* produce secretions that are released through ducts to nearby parts of the body. *Endocrine glands* produce *hormones* that are carried by the blood and lymph to all parts of the body.

II. Topics for Review

A. Classification of tissue
B. Functions of the four main types of tissues
C. Types of glands
D. Epithelial membranes

III. Matching Exercises

Matching only within each group, write the answers in the spaces provided.

Group A

cartilage	tissue	bone
adipose	cilia	squamous
exocrine	layered	

1. A group of cells similar in structure and function

   _____

2. Tiny hairlike projections from epithelium that can move dust and other foreign particles along the airways

   _____

3. The type of connective tissue that stores fat and serves as a heat insulator

   _____

4. The hard connective tissue that acts as a shock absorber and as a bearing surface to reduce friction between moving parts

   _____

5. Tissue that forms when cartilage gradually becomes impregnated with calcium salts

   _____

6. Term that describes flat, irregular epithelial cells

   _____

7. Another term that means *stratified*

   _____

8. Term for glands that secrete through ducts

   _____

Group B

ligament	cartilage	transitional
mucus	collagen	fascia

1. The secretion that traps dust and other inhaled foreign particles

   _____

2. A band or sheet of fibrous connective tissue around muscles

   _____

3. A wrinkled type of epithelium that is capable of great expansion

   _____

4. A strong connective band that supports a joint

   _____

5. The flexible white protein that makes up the main fibers in connective tissue

   _____

6. The tough, elastic substance found at the ends of long bones

   _____

Group C

myocardium	neuron	myelin
voluntary muscle	keloid	smooth muscle

1. Another name for a nerve cell _____

2. The fatty insulating material that covers and protects some nerve fibers _____

3. The thick, muscular layer of the heart wall _____

4. The result of excess production of collagen in the formation of a scar _____

5. Muscle tissue that forms the walls of the organs within the ventral body cavities _____

6. Term used to describe skeletal muscle because it is usually under the control of the will _____

Group D

suture   gland   capsule
epithelium   tendon   periosteum

1. The tissue that forms a protective covering for the body and that lines the intestinal tract and the respiratory and urinary passages _____

2. A layer of fibrous connective tissue around a bone _____

3. A specialized group of cells that manufactures substances from blood components _____

4. To bring the edges of a wound together to aid healing and reduce the size of a scar _____

5. A cord of connective tissue that connects a muscle to a bone _____

6. A tough connective tissue membrane that encloses an organ _____

Group E

pleura   membrane   serous membranes
pericardium   fascia   mucous membranes
peritoneum

1. Any thin sheet of material that separates two or more structures _____

2. The membranes that line the closed cavities within the body _____

3. A tough membrane composed entirely of connective tissue that serves to anchor and support an organ or to cover a muscle _____

4. The epithelial linings of tubes and spaces that are connected with the outside _____

5. The membrane that covers each lung _____

6. The special sac that encloses the heart _____

7. The large, serous membrane of the abdominal cavity _____

Group F

superficial fascia	periosteum	mucous membrane
parietal layer	perichondrium	cutaneous membrane
mesothelium	synovial membrane	

1. The tough connective tissue membrane that covers a bone _____

2. The membrane that covers cartilage _____

3. The lining of a joint cavity that produces secretions to reduce friction between the ends of bones _____

4. The type of lining found in the various parts of the respiratory tract _____

5. The tissue that underlies the skin _____

6. The part of a serous membrane that is attached to the wall of a cavity or sac _____

7. The type of epithelium that covers serous membranes _____

8. Another term for the skin _____

IV. Multiple Choice

Select the best answer and write the letter of your choice in the blank.

1. The study of tissues is   1. _____

   a. endocrinology
   b. cytology
   c. histology
   d. microbiology
   e. pharmacology

2. Which of the following is <u>not</u> a type of epithelial tissue?   2. _____

   a. transitional
   b. squamous
   c. cuboidal
   d. columnar
   e. areolar

3. The phrase *stratified squamous epithelium* describes

    a. flat, irregular, epithelial cells in many layers
    b. square epithelial cells in many layers
    c. long, narrow, epithelial cells in a single layer
    d. simple columnar epithelial cells
    e. flat, irregular, epithelial cells in a single layer

3. _____

4. Endocrine glands produce

    a. external secretions
    b. hormones
    c. digestive juices
    d. tears
    e. sweat

4. _____

5. Another name for areolar connective tissue is

    a. adipose tissue
    b. keloids
    c. fascia
    d. loose connective tissue
    e. voluntary muscle

5. _____

6. Adipose tissue stores

    a. mucus
    b. saliva
    c. fat
    d. cartilage
    e. bone

6. _____

V. Labeling

For each of the following illustrations, write the name or names of each labeled part on the numbered lines.

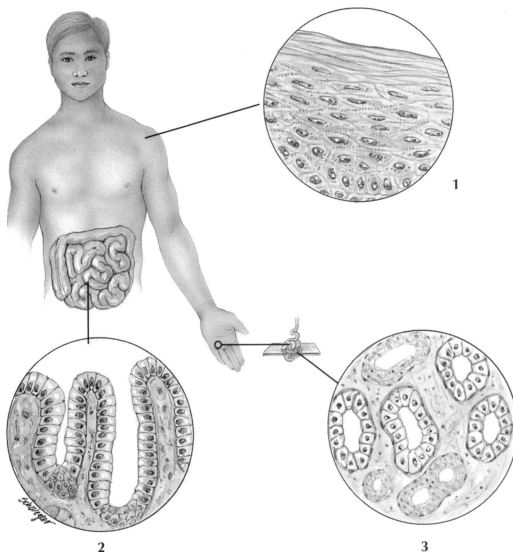

Three types of epithelium

1. _____   3. _____

2. _____

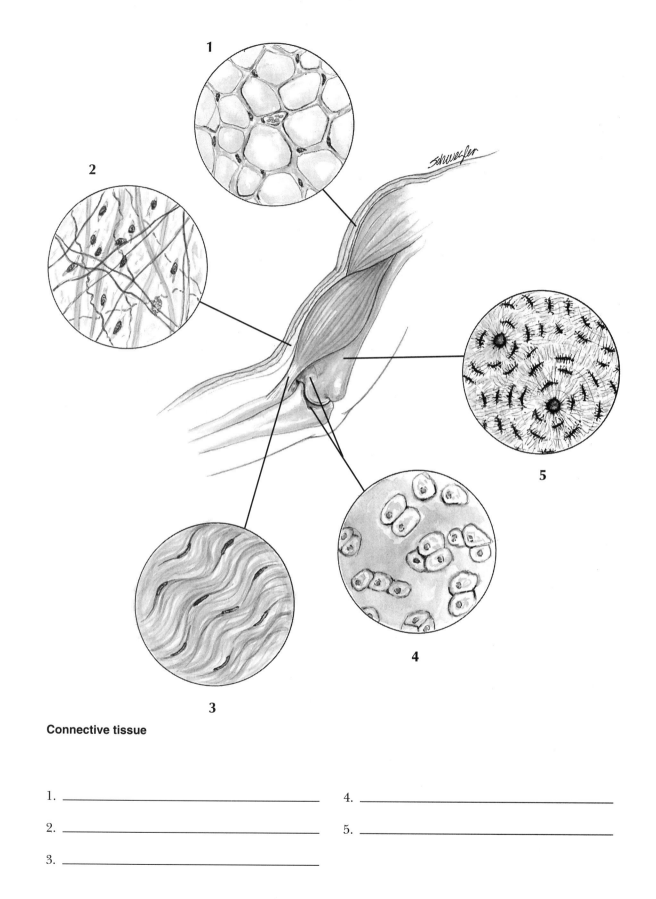

Connective tissue

1. _____   4. _____

2. _____   5. _____

3. _____

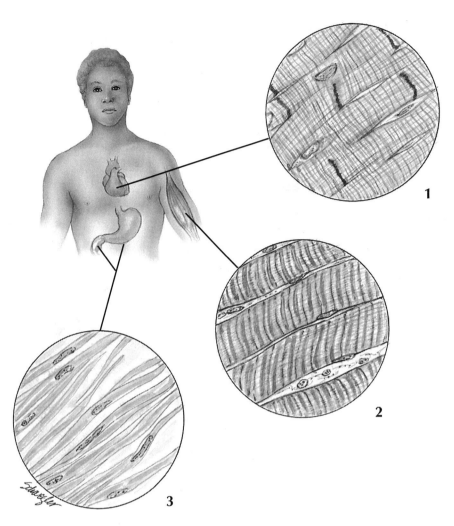

Muscle tissue

1. _____ 3. _____

2. _____

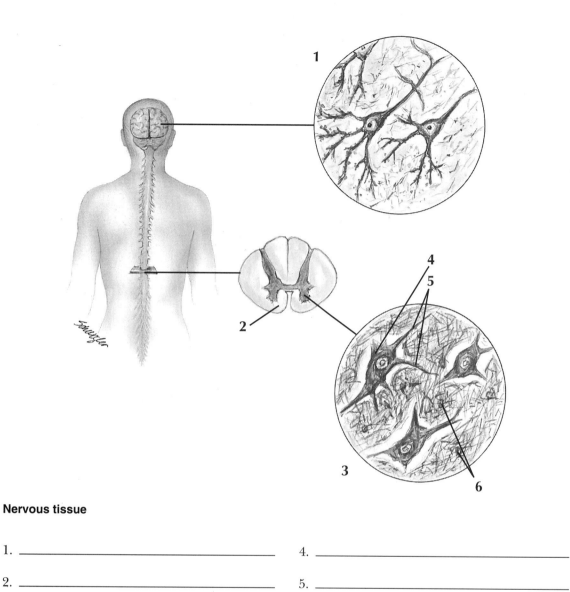

Nervous tissue

1. _____ 4. _____

2. _____ 5. _____

3. _____ 6. _____

VI. True-False

For each question, write T for true and F for false in the blank to the left of each number. If a statement is false, correct it by replacing the <u>underlined</u> term and write the correct statement in the blanks below the question.

_____ 1. The <u>exocrine</u> glands are ductless glands.

_____

_____

_____ 2. The <u>exocrine</u> glands produce external secretions.

_____

_____

_____ 3. Epithelium that is arranged in many layers is described as <u>simple</u>.

_____

_____

_____ 4. A <u>tendon</u> connects a bone to another bone.

_____

_____

_____ 5. <u>Periosteum</u> is the membrane around a bone.

_____

_____

_____ 6. <u>Smooth muscle</u> is also called visceral muscle.

_____

_____

_____ 7. The <u>axon</u> of a neuron carries impulses toward the cell body.

_____

_____

_____ 8. The <u>visceral</u> layer of a serous membrane lines the wall of a cavity or sac.

_____

_____

VII. Completion Exercise

Write the word or phrase that correctly completes each sentence.

1. Another term for the smooth involuntary muscle of most
   hollow organs is   _____

2. The supporting tissue of the body is called   _____

3. The basic unit of nervous tissue is the nerve cell, the scientific name for which is

_____

4. Movement is produced by the tissue known as

_____

5. The noun that means a serous membrane is

_____

6. Internal secretions, or hormones, affect tissues distant from the glands that produce them. These glands secrete directly into the bloodstream and are known as

_____

7. The lubricant produced by membranes that line cavities connected with the outside is known as

_____

8. The microscopic hairlike projections found in the cells lining most of the respiratory tract are called

_____

9. A layer of tough, fibrous connective tissue that encloses an internal organ is called a(n)

_____

10. The tough connective tissue membrane that covers most parts of all bones is given the name

_____

11. A lubricant that reduces friction between the ends of bones is produced by the

_____

VIII. Practical Applications

Study each discussion. Then write the appropriate word or phrase in the space provided.

Group A

The following cases were observed in an outpatient clinic.

1. Baby J experienced difficulty in breathing and had a copious discharge from his nose. A diagnosis of URI (upper respiratory infection) was made. The location of the membrane and the type of discharge indicated that the involved membrane was of a type known as a(n)

_____

2. Mrs. K had suffered a crushing injury to the lower leg. Initially she had little pain. Now she complains of numbness and pain in the foot and leg. This type of injury is made worse by the tight, fibrous covering of the muscles, known as the

_____

3. Mr. B was concerned about swelling and tenderness over his neck and upper back. His work involved the demolition of old buildings; he had become careless about personal cleanliness. Infection now involved the skin and connective tissue under it. The "sheet" that underlies the skin is called

_____

4. Mrs. C had undergone extensive surgery because of deformities due to rheumatoid arthritis, an inflammatory disorder of the membranes lining the joint spaces. These lining membranes are known as _____

5. Ms. G experienced abdominal pains following longstanding infection of the pelvic organs. Connective tissue bands were found to extend throughout the peritoneal surface. The layer of peritoneum that is attached to the organs is called the _____

6. Mrs. J was quite ill. Her symptoms were those associated with the disease called lupus erythematosus. She complained that it hurt to breathe because the membranes covering the lungs were involved. These membranes are called the _____

Group B

The following patients were seen in the hospital emergency room.

1. Little J, age 7, fell while riding his bicycle. He sustained several gashes on his face. At the emergency center the cuts were cleansed and sewn. The sewing of wounds to reduce the amount of scar tissue needed for healing is termed _____

2. Mrs. J had suffered a painful bump on her ankle. The swelling involved the superficial tissues and the fibrous covering of the bone, or the _____

3. Student N suffered a mild concussion while playing football, and it was feared that there might be damage to the brain coverings. These brain and spinal cord coverings are known as _____

IX. Short Essays

1. Define connective tissue and describe its role in the body, citing several examples of connective tissue.

   _____

   _____

   _____

   _____

2. Compare exocrine and endocrine glands with examples of each type.

_____

_____

_____

_____

3. Compare the three types of muscle tissue.

_____

_____

_____

_____

Memmler, RL, Cohen, BJ, Wood, DL. *STUDY GUIDE FOR STRUCTURE AND FUNCTION OF THE HUMAN BODY*, 6/e, © 1996, Lippincott-Raven Publishers

CHAPTER 5

The Skin

I. Overview

Because of its various properties, the skin can be classified as an *enveloping membrane,* an *organ,* and a *system.* A cross section of skin reveals its layers of *epidermis* (the outermost layer), *dermis* (the true skin where the skin glands are mainly located), and the *subcutaneous tissue* (the underlayer).

The skin *protects* deeper tissues against drying and against invasion by harmful organisms. It *regulates* body temperature through evaporation of sweat and loss of heat at the surface. It *obtains information* from the environment by means of sensory receptors.

Melanin is the main pigment that gives the skin its color. It functions to filter out harmful ultraviolet radiation from the sun. Races that have been exposed to the tropical sun for thousands of years have developed highly pigmented skin for protection. *Sebum,* secreted by the sebaceous glands, lubricates the skin and prevents dehydration. The protein *keratin* in the epidermis thickens and protects the skin. The hair and nails, structures associated with the skin, are composed mainly of keratin.

The appearance of the skin is influenced by such factors as the quantity of blood circulating in the surface blood vessels and its hemoglobin concentration. Aging, exposure to sunlight, and occupational activity also have a bearing on the condition and appearance of the skin.

As the most visible aspect of the body, the skin is the object of much quackery, and vast sums of money are spent in efforts to beautify it. Good general health is, however, the most important part of skin health and beauty.

II. Topics for Review

A. Skin layers
 1. Epidermis
 2. Dermis
 3. Subcutaneous layer
B. Skin glands
 1. Sudoriferous glands
 2. Sebaceous glands
C. Hair and nails
D. Functions of the skin
E. Observation of the skin

III. Matching Exercises

Matching only within each group, write the answers in the spaces provided.

Group A

epidermis	ciliary gland	integument
keratin	dermis	melanin

1. Another name for the skin as a whole _____

2. The protein in the epidermis that thickens and protects the skin _____

3. The outermost part of the skin, formed by several layers of epithelial cells _____

4. The true skin, or corium _____

5. The pigment that is largely responsible for skin color _____

6. A modified sweat gland found at the edge of the eyelid _____

Group B

sebaceous gland	subcutaneous tissue	sudoriferous gland
receptor	connective tissue	dermis

1. The tissue layer under the true skin _____

2. A sensory nerve ending in the skin that responds to information from the environment _____

3. The tissue that composes the framework of the dermis _____

4. A gland that produces sweat _____

5. The layer of the skin that contains most of the glands and hair _____

6. A gland that produces an oily secretion on skin and hair _____

Group C

melanin   stratum corneum   wax
nerve endings   sebum   follicle
stratum germinativum

1. The pigment that increases when the skin is exposed to sunlight   _____

2. The uppermost layer of the epidermis   _____

3. Secretion of the ceruminous glands   _____

4. The oily secretion of the sebaceous glands   _____

5. Structures in the skin that obtain information about the environment   _____

6. The deepest layer of the epidermis, which contains living, dividing cells   _____

7. The sheath in which a hair develops   _____

IV. Multiple Choice

Select the best answer and write the letter of your choice in the blank.

1. The subcutaneous layer of the skin is composed mainly of fibrous connective tissue and   1. _____

   a. nervous tissue
   b. cartilage
   c. epithelial tissue
   d. adipose tissue
   e. muscle tissue

2. Which of the following is <u>not</u> a function of skin?   3. _____

   a. protection
   b. breathing
   c. prevention of drying
   d. temperature regulation
   e. detection of changes in the environment

3. The body is cooled when blood vessels in the skin   4. _____

   a. dilate
   b. constrict
   c. become narrower
   d. close
   e. merge

V. Labeling

For each of the following illustrations, write the name or names of each labeled part on the numbered lines.

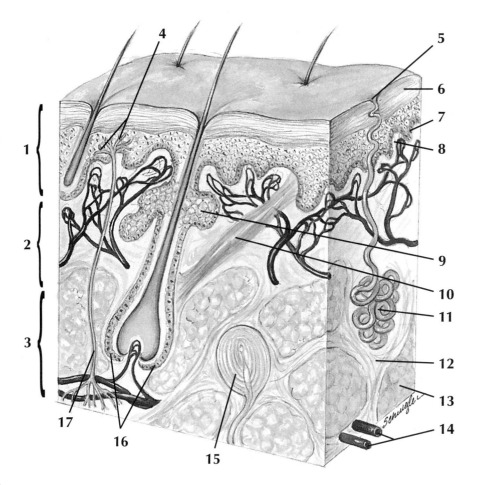

The skin

1. _____   10. _____

2. _____   11. _____

3. _____   12. _____

4. _____   13. _____

5. _____   14. _____

6. _____   15. _____

7. _____   16. _____

8. _____   17. _____

9. _____

VI. True-False

For each question, write T for true and F for false in the blank to the left of each number. If a statement is false, correct it by replacing the underlined term and write the correct statement in the blanks below the question.

_____ 1. The stratum germinativum is the uppermost layer of the epidermis.

_____

_____

_____ 2. When blood vessels enlarge they are said to dilate.

_____

_____

_____ 3. The subcutaneous tissue is also called the superficial fascia.

_____

_____

_____ 4. Sebum is produced by sudoriferous glands.

_____

_____

_____ 5. Melanin is a protein that helps to thicken the skin.

_____

_____

VII. Completion Exercise

Write the word or phrase that correctly completes each sentence.

1. The outer layer of the epidermis, the cells of which are constantly being shed, is designated the horny layer, or _____

2. The main pigment of the skin is _____

3. The ceruminous glands and the ciliary glands are modified forms of _____

4. Hair and nails are composed mainly of the protein that thickens and protects the skin. This protein is named _____

5. The blood vessels that nourish the epidermis are located in the skin layer just below the epidermis. This deeper layer is called the _____

VIII. Practical Applications

Study each discussion. Then print the appropriate word or phrase in the space provided.

1. L, age 15, consulted a skin doctor with his father. The son's skin was marked by pimples and blackheads and had a roughened appearance. This common disorder, call acne vulgaris, is found mainly in adolescents. It involves infection of the oil producing glands of the skin called the _____

2. Mr. M, a laborer, had neglected to give his skin proper care. He now had numerous painful boils in the axillae (armpits). These boils were caused by the entrance of bacteria into the sheaths in which the hairs grow. These sheaths are called _____

3. There was also a deep-seated infection of the tissue under the skin on Mr. M's lower back. This tissue lying under the skin is described by the term _____

4. Mt. Laurel Hospital used footprints to identify all newborn babies in the nursery. The pattern of a footprint is formed by elevations and depressions in the skin layer beneath the epidermis, the layer called the _____

5. Ms. J, age 17, consulted her family physician because she had noticed scattered dark areas on the surface of her skin. The doctor thought her problem was due simply to too much sun exposure. The skin pigment that gives the skin color and increases when one is exposed to sunlight is _____

6. Mr. G's teenage son had large areas of red, peeling skin due to sunburn. The nurse taught him skin care, including how to prevent injury to the skin by applications of _____

7. The clinic personnel were given careful instructions on the proper method of handwashing. This is desirable because handwashing is the most effective method of preventing the spread of _____

IX. Short Essays

1. Explain why the skin is classified in several different ways.

   _____

   _____

   _____

   _____

2. Name several pigments that can give color to skin and identify their sources.

   _____

   _____

   _____

   _____

3. Describe the changes that may occur in skin with aging.

   _____

   _____

   _____

   _____

Movement and Support

6. THE SKELETON: BONES AND JOINTS

7. THE MUSCULAR SYSTEM

Memmler, RL, Cohen, BJ, Wood, DL. *STUDY GUIDE FOR STRUCTURE AND FUNCTION OF THE HUMAN BODY*, 6/e, © 1996, Lippincott-Raven Publishers

CHAPTER 6

The Skeleton: Bones and Joints

I. Overview

The skeletal system protects and supports the body parts and serves as an attachment for the muscles, which furnish the power for movement. The skeletal system includes some 206 bones; the number varies slightly according to age and the individual.

Although bone tissue contains a large proportion of nonliving material, bones also contain living cells and have their own systems of blood vessels, lymphatic vessels, and nerves. Bone tissue may be either *spongy* or *compact.* Compact bone is found in the shaft (diaphysis) of a long bone and in the outer layer of other bones. Spongy bone makes up the ends (epiphyses) of a long bone and the center of other bones. *Red marrow* occurs in certain parts of all bones and manufactures the blood cells; *yellow marrow,* which is largely fat, is found in the central cavities of the long bones.

Bone tissue is produced by cells called *osteoblasts,* which gradually convert cartilage to bone during development. The mature cells that maintain bone are called *osteocytes,* and the cells that break down bone for remodeling and repair are the *osteoclasts.*

The entire bony framework of the body is called the skeleton. It is divided into two main groups of bones, the *axial skeleton* and the *appendicular skeleton.* The axial skeleton includes the skull, spinal column, ribs, and sternum. The appendicular skeleton consists of the bones of the arms and legs, the shoulder girdle, and the pelvic girdle.

A *joint* is the region of union of two or more bones; joints are classified according to structure and to the degree of movement permitted. *Synovial joints* show the greatest degree of movement. The six types of synovial joints allow for a variety of movements in different directions. Connective tissue bands, *ligaments,* hold the bones together in all the synovial joints and many of the less movable joints.

II. Topics for Review

A. Structure of bone
B. Bone cells
C. Bone growth and repair
D. Functions of bones
E. Bone markings
F. Bones of the axial skeleton
G. Bones of the appendicular skeleton
H. Types of joints
 I. Movements at synovial joints

III. Matching Exercises

Matching only within each group, write the answers in the spaces provided.

Group A

cartilage   red marrow   appendicular skeleton
periosteum   endosteum   osteoblast
axial skeleton   yellow marrow

1. The bony framework of the head and trunk together _____

2. The site of blood cell production _____

3. The combination of bones that form the framework for the extremities _____

4. The fatty material found inside the central cavities of long bones _____

5. The tough connective tissue membrane that covers bones _____

6. The somewhat thinner membrane that lines the central cavity of long bones _____

7. The material that forms the skeleton in the embryo _____

8. A cell that produces bone _____

Group B

frontal bone   temporal bones   parietal bones
ethmoid bone   occipital bone   sphenoid bone

1. The bone located between the eyes that extends into the nasal cavity, eye sockets, and cranial floor _____

2. The bone that forms the back of the skull and part of the base of the skull

_____

3. The bone that forms the forehead

_____

4. The bat-shaped bone that extends behind the eyes and also forms part of the base of the skull

_____

5. The paired bones that form the larger part of the upper and side walls of the cranium

_____

6. The two bones that form the lower sides and part of the base of the central areas of the skull

_____

Group C

mandible	maxillae	zygomatic bone
hyoid	nasal bones	lacrimal bone

1. The very small bone at the inside corner of the eye

_____

2. The only movable bone of the skull

_____

3. The U-shaped bone lying just below the skull proper

_____

4. The bone that forms the upper part of the cheek

_____

5. The two bones of the upper jaw

_____

6. The two slender bones that form much of the bridge of the nose

_____

Group D

true ribs	diaphysis	floating ribs
lumbar region	coccyx	thoracic region
cervical region		

1. The shaft of a long bone

_____

2. The region of the spinal column, made up of the first seven vertebrae, comprising the main framework of the neck

_____

3. The third section of the vertebral column, consisting of five large vertebrae

_____

4. The second part of the vertebral column, made up of 12 vertebrae

_____

5. The tail part of the vertebral column, made of four or five small fused bones

_____

6. Term for the first seven pairs of ribs as a group

_____

7. The last two pairs of false ribs, which are very short and do not extend to the front of the body

_____

Group E

patella	fibula	olecranon
ulna	tibia	radius

1. The upper part of the ulna, which forms the point of the elbow

_____

2. The medial forearm bone

_____

3. The scientific name for the kneecap

_____

4. The larger of the two leg bones

_____

5. The bone located on the thumb side of the forearm

_____

6. The lateral bone of the leg

_____

Group F

costal	foramina	epiphysis
spongy bone	fontanelle	sacrum

1. A soft spot in the infant skull that later closes

_____

2. The end of a long bone

_____

3. The type of bone tissue found at the ends of a long bone

_____

4. The region of the spinal column below the lumbar region, composed of four to five fused bones

_____

5. An adjective that refers to the ribs

_____

6. Openings or holes that extend into or through bones

_____

Group G

greater trochanter	clavicle	carpal bones
metacarpal bones	calcaneus	ilium
ligament	phalanges	

1. The five bones in the palm of the hand

2. The largest of the tarsal bones; the heel bone

3. The 14 small bones that form the framework of the fingers on each hand

4. The upper wing-shaped part of the os coxae in the pelvic girdle

5. The bones of the wrist

6. A bone of the shoulder girdle

7. A connective tissue band that holds bones together at a joint

8. The large, rounded projection at the upper and lateral portion of the femur

Group H

articular cartilage	ball-and-socket	abduction
synovial	flexion	hinge

1. The protective layer of tissue that covers the contacting bone surfaces at a joint

2. A bending motion that decreases the angle between bones

3. Term for the lubricating fluid inside a joint cavity

4. Movement away from the midline of the body

5. The type of joint that allows for circumduction

6. The type of joint found at the elbow

Group I

articulation	acetabulum	rotation
diaphysis	extension	diarthrosis

1. The deep socket in the hip bone that holds the head of the femur

2. The shaft of a long bone

3. A motion that increases the angle between bones _____

4. The region of union of two or more bones; a joint _____

5. A freely movable joint _____

6. Motion around a central axis _____

IV. Multiple Choice

Select the best answer and write the letter of your choice in the blank.

1. The hard and brittle material in bones is composed mainly of salts of the element

   a. iodine
   b. sodium
   c. chlorine
   d. calcium
   e. nitrogen

   1. _____

2. The part of the skull that encloses the brain is the

   a. hyoid bone
   b. cranium
   c. sinus
   d. conchae
   e. vomer

   2. _____

3. A joint between bones of the skull is a

   a. cleft palate
   b. suture
   c. trochanter
   d. malleolus
   e. crest

   3. _____

4. The patella is the largest of a type of bone that develops within a tendon or a joint capsule. It is described as

   a. synovial
   b. axial
   c. rheumatoid
   d. symphysis
   e. sesamoid

   4. _____

5. The lower end of the sternum, used as a landmark for cardiopulmonary resuscitation, is the

   5. _____

   a. xiphoid process
   b. ischial spine
   c. lateral malleolus
   d. acromion
   e. glenoid cavity

6. The foramen magnum is

   6. _____

   a. a large hole in a hip bone near the symphysis pubis
   b. the curved rim along the top of the hip bone
   c. a hole between vertebrae that allows for passage of a spinal nerve
   d. a process on the temporal bone
   e. a large opening at the base of the skull through which the spinal cord passes

V. Labeling

For each of the following illustrations, write the name or names of each labeled part on the numbered lines.

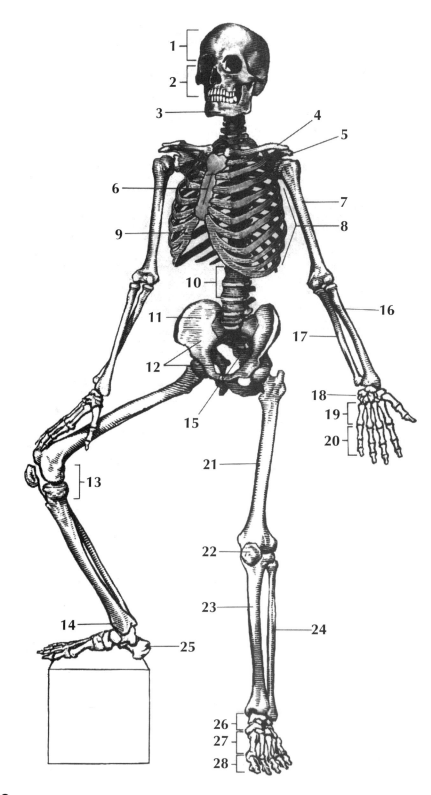

The skeleton

1. _____

2. _____

3. _____

4. _____

5. _____

6. _____

7. _____

8. _____

9. _____

10. _____

11. _____

12. _____

13. _____

14. _____

15. _____

16. _____

17. _____

18. _____

19. _____

20. _____

21. _____

22. _____

23. _____

24. _____

25. _____

26. _____

27. _____

28. _____

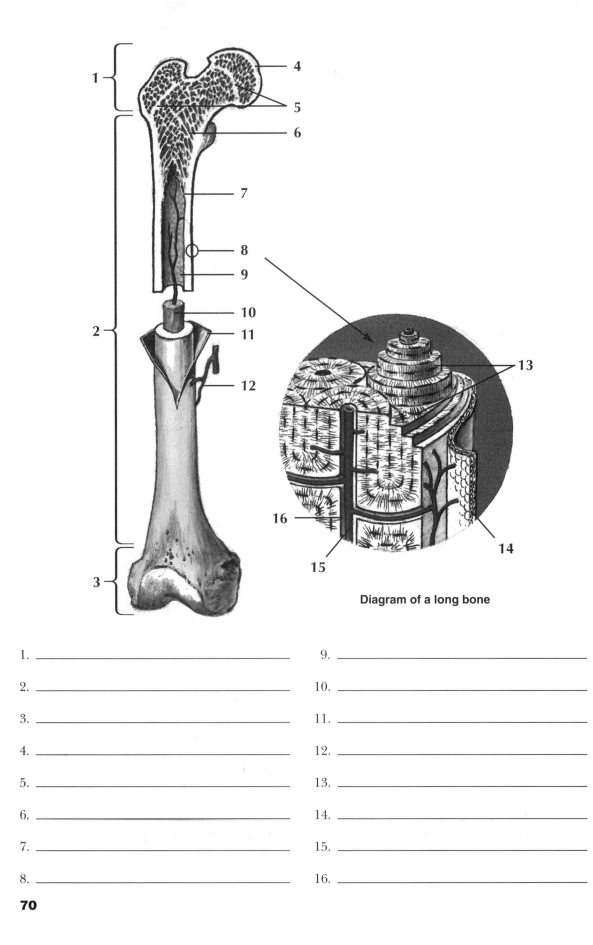

Diagram of a long bone

1. _____
2. _____
3. _____
4. _____
5. _____
6. _____
7. _____
8. _____

9. _____
10. _____
11. _____
12. _____
13. _____
14. _____
15. _____
16. _____

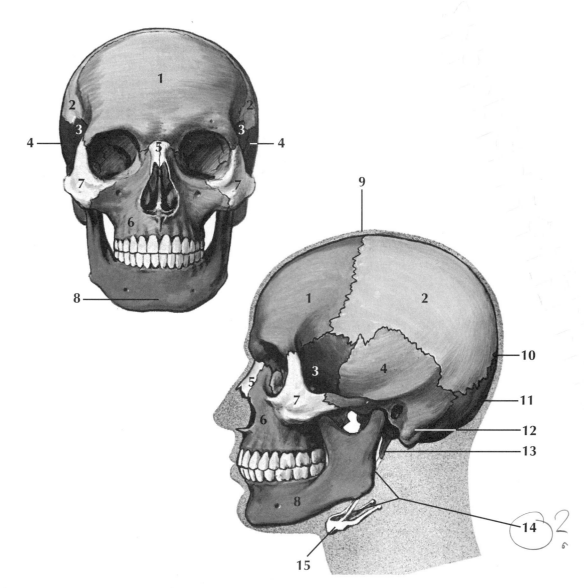

Skull from the front and from the left

1. Frontal
2. parietal
3. Sphenoid
4. temporal
5. Nasal bone
6. maxilla
7. zygomatic
8. mandible

9. coronoid suture
10. sagittal suture
11. occipital bone
12. mastoid process
13. Ramus
14.
15. hyoid bone

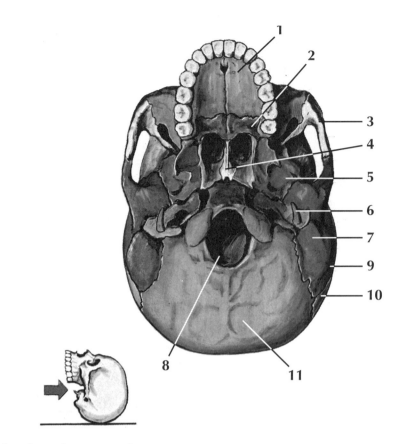

Skull from below, lower jaw removed

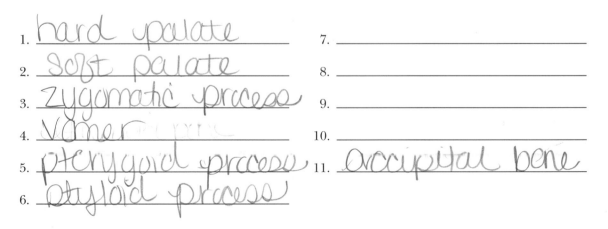

1. hard palate
2. soft palate
3. zygomatic process
4. vomer
5. pterygoid process
6. styloid process
7. _____
8. _____
9. _____
10. _____
11. occipital bone

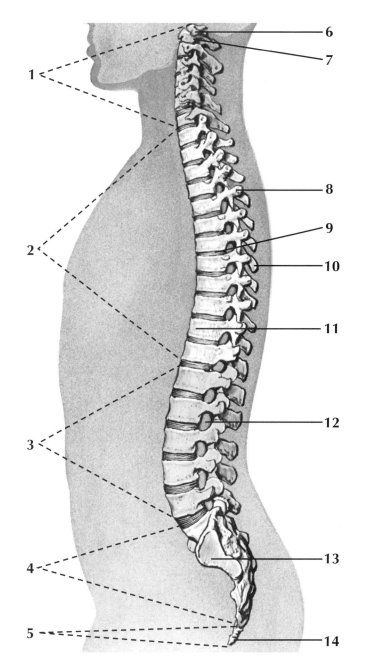

Vertebral column

1.	_____
2.	_____
3.	_____
4.	_____
5.	_____
6.	_____
7.	_____
8.	_____
9.	_____
10.	_____
11.	_____
12.	_____
13.	_____
14.	_____

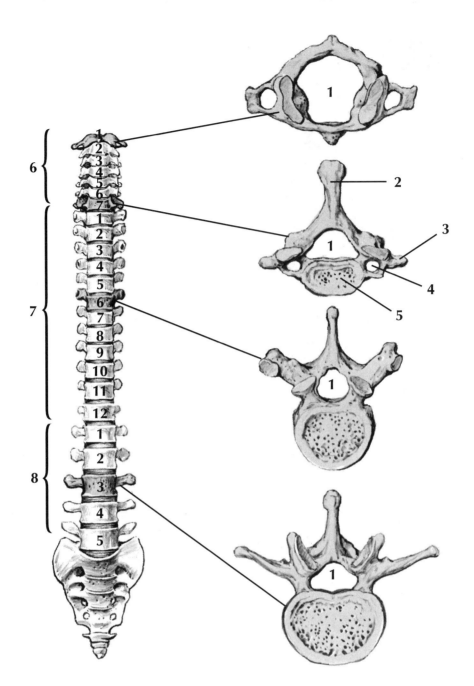

Vertebrae

1. _____   5. _____

2. _____   6. _____

3. _____   7. _____

4. _____   8. _____

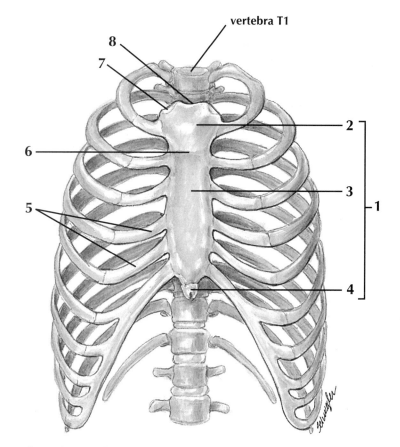

Bones of the thorax (anterior view)

vertebra T1

8

7

2

6

3

5

1

4

1. _____   5. _____

2. _____   6. _____

3. _____   7. _____

4. _____   8. _____

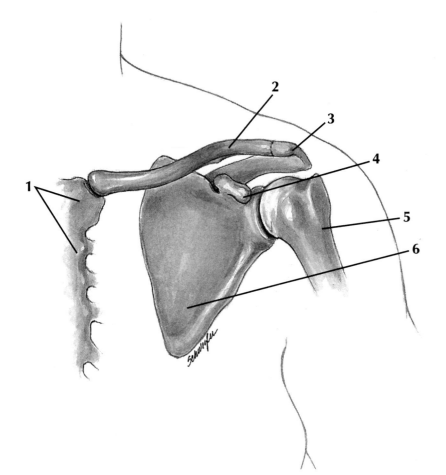

Bones of the shoulder girdle

1. _____   4. _____

2. _____   5. _____

3. _____   6. _____

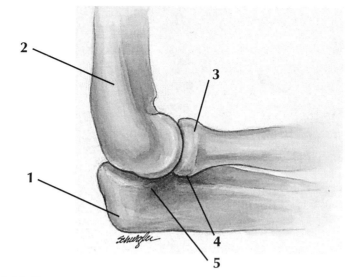

Lateral view of the right elbow

1. _____   4. _____

2. _____   5. _____

3. _____

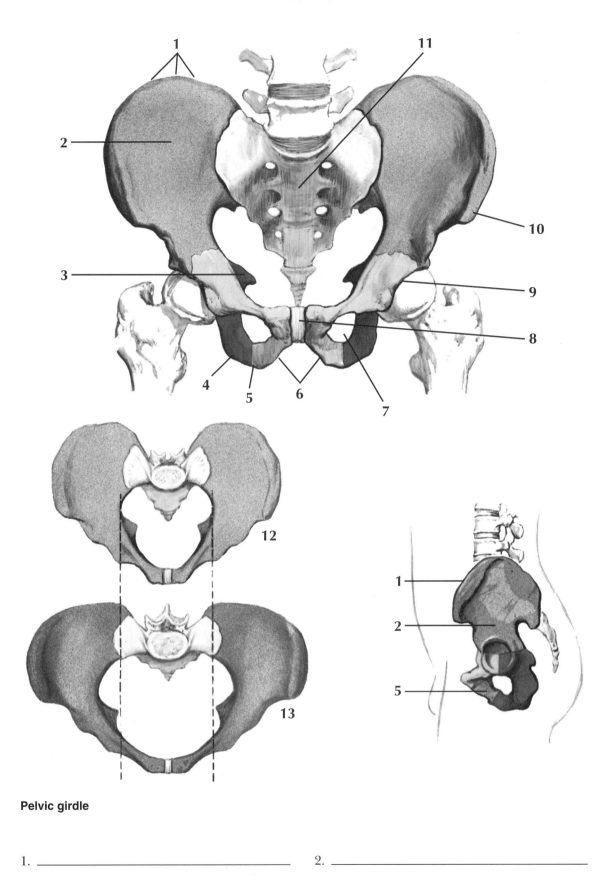

Pelvic girdle

1. _____ 2. _____

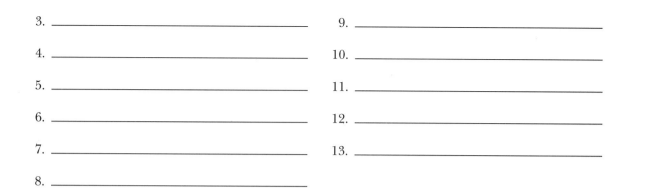

3. _____   9. _____

4. _____   10. _____

5. _____   11. _____

6. _____   12. _____

7. _____   13. _____

8. _____

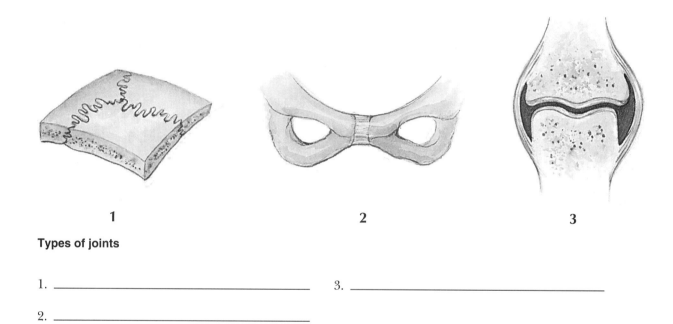

| 1 | 2 | 3 |

Types of joints

1. _____   3. _____

2. _____

VI. True-False

For each question, write T for true and F for false in the blank to the left of each number. If a statement is false, correct it by replacing the underlined term and write the correct statement in the blanks below the question.

_____ 1. The zygomatic bone is part of the underlined appendicular skeleton.

_____

_____

_____ 2. The patella is part of the appendicular skeleton.

_____

_____

_____ 3. There are <u>seven</u> pairs of true ribs.

_____

_____

_____ 4. In the anatomic position, the radius is <u>medial</u> to the ulna.

_____

_____

_____ 5. The shaft of a long bone is the <u>epiphysis</u>.

_____

_____

_____ 6. In a newborn infant, the entire vertebral column is <u>concave</u>.

_____

_____

_____ 7. The palm is turned up or forward in <u>pronation</u>.

_____

_____

_____ 8. During development, transformation of cartilage into bone begins at the center of the <u>diaphysis</u>.

_____

_____

_____ 9. Movement of a part away from the midline of the body is termed <u>adduction</u>.

_____

_____

VII. Completion Exercise

Write the word or phrase that correctly completes each sentence.

1. When bone-forming cells mature and become enclosed in hardened bone material, they are referred to as _____

2. In the embryonic stage of bone development, most of the developing bones are made of _____

3. When bone is resorbed, cells that break down bone become active; these cells are called _____

4. The type of bone tissue that makes up the shaft of a long bone is called _____

5. The skull, vertebrae, ribs, and sternum make up the division of the skeleton called the _____

6. The cervical and lumbar curves, which appear after birth, are referred to as _____

7. A suture is an example of an immovable joint also called a(n) _____

8. Pivot, hinge, and gliding joints are examples of freely movable joints also called _____

9. Winding up to throw a pitch requires a broad circular movement at the shoulder that is a combination of simpler movements. This combined motion is called _____

10. When a toe dancer points her toes downward and flexes the arch of her foot, the motion is technically called _____

VIII. Practical Applications

Study each discussion. Then write the appropriate word or phrase in the space provided.

Group A

A group of high-school seniors was involved in a serious traffic accident on the way home from the prom.

1. There was a pronounced swelling of the upper right side of Michelle's head. X-ray films showed a fracture of the bone that forms the top and side of the cranium. This bone is the _____

2. Michelle also suffered an injury to one of the two large bones of the pelvic girdle. This bone articulates with the sacrum and is named the _____

3. Jason suffered multiple injuries to his left lower extremity. Protruding through the skin was a splintered portion of the longest bone in the body, the _____

4. Susan thought her injuries were the least serious, so she walked several blocks to find help. Then she noticed that her right knee was not functioning normally. Examination revealed a fractured kneecap. The scientific name for the kneecap is _____

5. Glen, the driver of the car, was forcibly thrown against the steering wheel. He suffered fractures of the sixth and seventh ribs, which are the last ribs in the group called the _____

Group B

Mr. B, age 58, was admitted to the general hospital because of acute pain and swelling of his right great toe. He also complained of a chronic backache. Mr. B underwent a complete physical examination.

1. X-ray films showed involvement of the toe joints. The framework of the toes is made up of bones called the _____

2. Spurs of bony material were found to be present at the edges of the vertebrae just above the sacrum. These bones form the region of the spinal column called the _____

Group C

Mrs. C, age 36, visited her doctor's office because of swelling and pain in the joints of her hands and fingers. Examination revealed the following:

1. Evidence of inflammation and overgrowth of the membrane lining the joint cavities, a membrane that is called the _____

2. Difficulty in moving the joints of the fingers due to damage to the normally smooth tissue on the joint surface. This layer is called the _____

IX. Short Essays

1. Explain why bone is described as living tissue.

   _____

   _____

   _____

   _____

2. Describe the four curves of the adult spine and explain the purpose of these curves

   _____

   _____

   _____

   _____

3. Describe the structures that strengthen and stabilize a freely movable joint.

_____

_____

_____

_____

Memmler, RL, Cohen, BJ, Wood, DL. *STUDY GUIDE FOR STRUCTURE AND FUNCTION OF THE HUMAN BODY*, 6/e, © 1996, Lippincott-Raven Publishers

CHAPTER 7

The Muscular System

I. Overview

There are three basic types of muscle tissue: ***skeletal, smooth,*** and ***cardiac.*** The focus of this chapter is skeletal muscle, which is attached to bones. The muscular system is composed of more than 650 individual muscles. Muscles usually work in groups to execute a body movement. The muscle that produces a given movement is called the ***prime mover;*** the muscle that produces the opposite action is the ***antagonist.*** Skeletal muscle is also called voluntary muscle, because normally it is under conscious control.

Skeletal muscles are activated by electric impulses from the nervous system. A nerve fiber makes contact with a muscle cell at the ***neuromuscular junction.*** From this point, the impulse spreads along the muscle cell membrane, producing an electric change called the ***action potential.*** As a result of this electric change in the cells, the muscle can contract (shorten) to produce movement.

Muscle contraction occurs by the sliding together of protein filaments called ***actin*** and ***myosin*** within the cell. These filaments make contact only in the presence of calcium, which is released from the endoplasmic reticulum of the muscle cell when the action potential spreads along the cell membrane. ***ATP*** is the direct source of energy for the contraction. To manufacture ATP, the cell must have adequate supplies of glucose and oxygen delivered by the blood. A reserve supply of glucose is stored in muscle cells in the form of a compound called ***glycogen,*** and additional oxygen is stored by a pigment in the cells called ***myoglobin.***

When muscles do not receive enough oxygen, as during strenuous activity, they can produce a small amount of ATP and continue to function for a short period. As a result, however, the cells produce lactic acid, which eventually causes muscle fatigue. The individual must then rest and continue to breathe in oxygen, which is used to convert the lactic acid into other substances. The amount of oxygen needed for this purpose is referred to as the *oxygen debt.*

Muscles act with the bones of the skeleton as *lever systems,* in which the joint is the pivot point or *fulcrum.* Exercise and proper body mechanics help in maintaining muscle health and effectiveness. Continued activity delays the undesirable effects of aging.

II. Topics for Review

A. General characteristics of skeletal muscles
B. The mechanism of muscle contraction
C. The energy for muscle contraction
D. Effects of exercise
E. Muscle movement
F. Muscle attachments
G. Body mechanics
H. Muscles of the head and the neck
 I. Muscles of the upper extremities
 J. Muscles of the trunk
K. Muscles of the lower extremities

III. Matching Exercises

Matching only within each group, write the answers in the spaces provided.

Group A

action potential   contractility   irritability
neuromuscular junction   isotonic   tonus
isometric

1. The capacity of a muscle to respond to a stimulus   _____

2. The point where a motor nerve fiber contacts a muscle cell   _____

3. The electrical change transmitted along the muscle cell membrane after stimulation   _____

4. The capacity of a muscle fiber to undergo shortening   _____

5. The normal partially contracted state of muscles   _____

6. Term applied to muscle contractions in which the tone remains constant while the muscle shortens   _____

7. Term applied to contractions in which there is a great increase in muscle tension without change in muscle length   _____

Group B

glycogen   actin   lactic acid
calcium   myoglobin   ATP

1. The substance that accumulates in muscles working without enough oxygen

_____

2. The ion that must be released into the muscle cell before contraction

_____

3. The immediate source of energy for muscle contraction

_____

4. The compound that stores glucose in muscle cells

_____

5. A protein filament needed to produce contraction in muscle cells

_____

6. The compound that stores oxygen in muscle cells

_____

Group C

vasodilation   myosin   prime mover
antagonist   origin   insertion

1. The muscle that produces a given movement

_____

2. The muscle attachment joined to a moving part of the body

_____

3. A widening of a blood vessel

_____

4. The muscle attachment joined to a more fixed part of the body

_____

5. A protein needed for contraction in muscle cells

_____

6. Name for the muscle that must relax during a given movement

_____

Group D

biceps brachii   pectoralis major   deltoid
latissimus dorsi   triceps brachii   trapezius
sternocleidomastoid

1. One of the two muscles on either side of the neck that flexes the head on the chest

_____

2. A triangular muscle over the back and neck that moves the shoulder

_____

3. The muscle of the middle and lower back that is a powerful extensor of the arm (at the shoulder)

_____

4. The muscle capping the shoulder and upper arm

_____

5. A muscle on the front of the arm that acts as a flexor of the elbow and a supinator of the hand

_____

6. The large muscle on the back of the arm that extends the elbow, as when delivering a blow

   _____

7. The large muscle of the upper chest that flexes the arm across the body

   _____

Group E

gastrocnemius   levator ani   deep fascia
aponeurosis   tendon   diaphragm
torticollis   rotator cuff

1. A cordlike structure that attaches a muscle to bone

   _____

2. A sheet of connective tissue that attaches certain muscles to bone or other muscles

   _____

3. A connective tissue sheath enclosing an entire muscle

   _____

4. The chief muscle of respiration

   _____

5. The chief muscle of the calf of the leg

   _____

6. The muscle of the pelvic floor that aids in defecation

   _____

7. A condition that may be caused by injury or spasm of a sternomastoid muscle

   _____

8. A muscle group that supports the shoulder joint

   _____

Group F

sacrospinalis   intercostal   gluteus maximus
buccinator   sartorius   quadriceps femoris
iliopsoas

1. Muscles that aid in respiration and are located between the ribs

   _____

2. The muscle that forms the fleshy part of the cheek

   _____

3. The longest muscle of the spine

   _____

4. The muscle that forms much of the fleshy part of the buttock

   _____

5. The powerful flexor of the thigh

   _____

6. The muscle that extends the knee, as in kicking a ball

   _____

7. The thin muscle that travels down and across the medial surface of the thigh

   _____

IV. Multiple Choice

Select the best answer and write the letter of your choice in the blank.

1. The lateral muscle of the leg that turns the sole of the foot outward (eversion) is the

   a. peroneus longus
   b. internal oblique
   c. extensor carpi
   d. teres minor
   e. adductor longus

   1. _____

2. Which of the following statements is <u>not</u> true of skeletal muscle?

   a. The cells are long and threadlike.
   b. It is normally under conscious control.
   c. It is described as striated.
   d. The cells are multinucleated
   e. It is involuntary.

   2. _____

3. When muscles and bones act together in the body as a lever system, the pivot point or fulcrum of the system is the

   a. joint
   b. tendon
   c. extensor
   d. myoglobin
   e. levator

   3. _____

4. The hamstring muscles act to

   a. extend the leg
   b. flex the leg
   c. abduct the thigh
   d. adduct the arm
   e. move the hand

   4. _____

V. Labeling

For each of the following illustrations, write the name or names of each labeled part on the numbered lines.

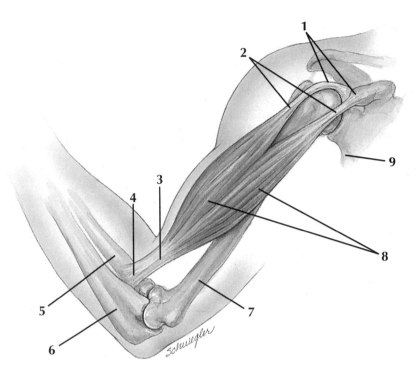

Diagram of a muscle

1. _____   6. _____

2. _____   7. _____

3. _____   8. _____

4. _____   9. _____

5. _____

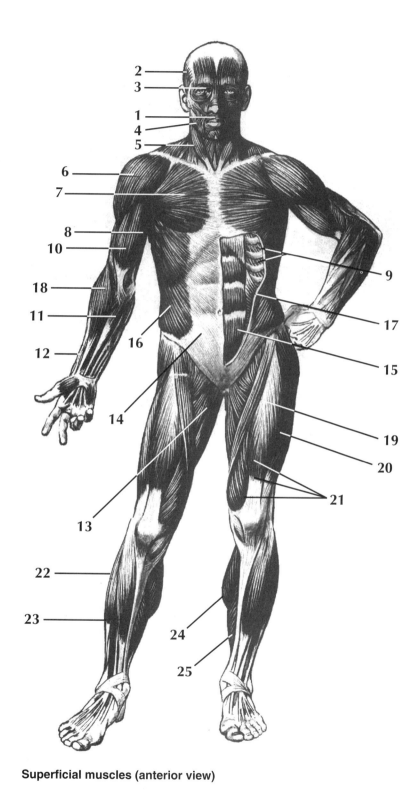

Superficial muscles (anterior view)

1. _____
2. _____
3. _____
4. _____
5. _____
6. _____
7. _____
8. _____
9. _____
10. _____
11. _____
12. _____
13. _____
14. _____
15. _____
16. _____
17. _____
18. _____
19. _____
20. _____
21. _____
22. _____
23. _____
24. _____
25. _____

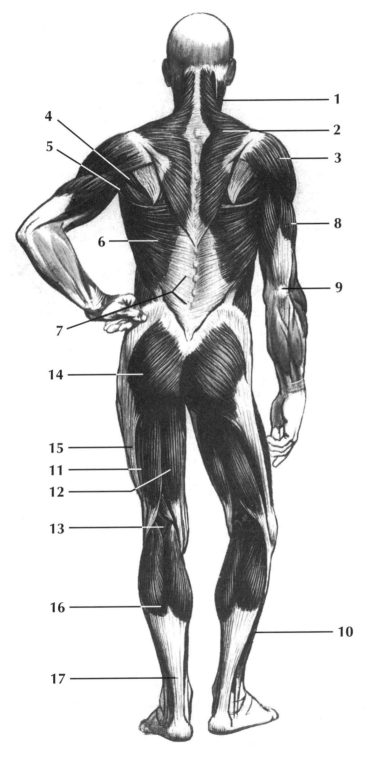

1. _____

2. _____

3. _____

4. _____

5. _____

6. _____

7. _____

8. _____

9. _____

10. _____

11. _____

12. _____

13. _____

14. _____

15. _____

16. _____

17. _____

Superficial muscles (posterior view)

Location of diaphragm

1. _____   4. _____

2. _____   5. _____

3. _____

VI. True-False

For each question, write T for true and F for false in the blank to the left of each number. If a statement is false, correct it by replacing the underlined term and write the correct statement in the blanks below the question.

_____ 1. In an isotonic contraction, muscle tension increases but the muscle does not shorten.

_____

_____

_____ 2. Muscles enter into oxygen debt when they are functioning anaerobically.

_____

_____

_____ 3. <u>Actin</u> is the light filament in skeletal muscle cells.

_____

_____

_____ 4. The <u>origin</u> of a muscle is attached to a moving part of the body.

_____

_____

_____ 5. The biceps brachii <u>extends</u> the arm at the elbow.

_____

_____

_____ 6. The intercostal muscles are between the <u>ribs</u>.

_____

_____

VII. Completion Exercise

Write the word or phrase that correctly completes each sentence.

1. Normally, muscles are in a partially contracted state, even though they are not in use at the time. This state of mild constant tension is called _____

2. A movement is initiated by a muscle or set of muscles called the _____

3. The end of a muscle that is attached to a part moved by that muscle is the _____

4. The muscle of the lips is the _____

5. Muscles functioning without enough oxygen will fatigue as a result of the accumulation of _____

6. The muscle attachment that is usually relatively fixed is called its _____

7. The movement of a prime mover is opposed by a muscle or set of muscles called the _____

8. A group of muscles that covers the front and sides of the femur and extends the leg is the _____

9. There are four pairs of muscles for chewing. The muscle located at the angle of the jaw is called the

_____

10. A superficial muscle of the neck and upper back acts on the shoulder. This muscle is the

_____

11. The muscle on the front of the leg that raises the sole of the foot (dorsiflexion) is the

_____

12. The muscular partition between the thoracic and abdominal cavities is the

_____

13. The band of connective tissue that attaches the gastrocnemius muscle to the heel is the

_____

VIII. Practical Applications

Study each discussion. Then print the appropriate word or phrase in the space provided.

Group A

Driver J and his three companions tried to race an oncoming train to an intersection. J misjudged the speed of the train, and the train crashed into the car. All four occupants of the car received multiple injuries.

1. Driver J was thrown against the steering wheel, which punctured his chest. This puncture involved the muscles between the ribs, called the

_____

2. Mr. K, the occupant sitting next to the driver, suffered facial injuries, in which the muscle that encircles the eye was cut. This muscle is called the

_____

3. Ms. L was thrown out of the car and received lacerations and fractures of the lower extremities, including the calf of the leg. The largest muscle of the leg is the

_____

4. Mr. M received shoulder and upper back lacerations. They involved the muscle that covers the shoulder and abducts the arm, the

_____

Group B

In the physical therapy department, several patients were receiving physical therapy for muscle injuries.

1. Mrs. K had suffered a stroke that involved the left lower extremity. One of the large muscles used in standing forms most of the buttock, and is named the

_____

2. Mr. P had suffered a fracture of the humerus and was receiving treatment for the damage to the large extensor of the elbow, located on the dorsal part of the arm. This muscle is the _____

3. Ms. L had been in a cast for a number of weeks, so she was receiving exercises for the strengthening of many body muscles, including the large muscle that originates from the middle and lower back and inserts on the arm bone (humerus). This strong swimming muscle is the _____

4. Ms. R, age 76, came in for exercises to strengthen some of her extensor muscles to prevent the further deterioration of her spine. The long extensor muscle of the back, which helps maintain an erect posture, needed particular attention. This is the _____

5. Mr. J, a weekend athlete, was seen for exercises after surgical repair of a group of deep muscles of the shoulder. This group of muscles allows for the wide range of movement of the shoulder and is known as the _____

IX. Short Essays

1. Name and compare the three types of muscle tissue.

   _____

   _____

   _____

   _____

2. List the requirements for skeletal muscle contraction and explain the role of each.

   _____

   _____

   _____

   _____

3. Briefly describe what happens in muscle cells functioning anaerobically.

   _____

   _____

   _____

   _____

III

UNIT

Coordination and Control

8. THE NERVOUS SYSTEM: THE SPINAL CORD AND SPINAL NERVES

9. THE NERVOUS SYSTEM: THE BRAIN AND CRANIAL NERVES

10. THE SENSORY SYSTEM

11. THE ENDOCRINE SYSTEM: GLANDS AND HORMONES

Memmler, RL, Cohen, BJ, Wood, DL. *STUDY GUIDE FOR STRUCTURE AND FUNCTION OF THE HUMAN BODY*, 6/e, © 1996, Lippincott-Raven Publishers

CHAPTER 8

The Nervous System: The Spinal Cord and Spinal Nerves

I. Overview

The nervous system is the body's coordinating system—receiving, sorting, and controlling responses to both internal and external stimuli. This system functions by means of the *nerve impulse,* an electric current that spreads along the membrane of the nerve cell or *neuron.* Each neuron is composed of a cell body and nerve fibers, which are threadlike extensions from the cell body. A *dendrite* is a fiber that carries impulses toward the cell body, and an *axon* is a fiber that carries impulses away from the cell body. Some axons are covered with a sheath of fatty material called *myelin,* which insulates the fiber and speeds conduction along the fiber. Nerve fibers are collected in bundles to form *nerves.*

Nerve cells make contact at a junction called a *synapse.* Here, the nerve impulse travels across a very narrow cleft between the cells by means of a chemical referred to as a *neurotransmitter.* Neurotransmitters are released from axons to be picked up by receptors in the membranes of responding cells.

A neuron may be classified as either a sensory (afferent) type, which carries impulses toward the central nervous system, or a motor (efferent) type, which carries impulses away from the central nervous system. There are also connecting neurons within the central nervous system. The basic functional pathway of the nervous system is a *reflex arc,* in which an impulse travels from a receptor, along a sensory neuron to a synapse or synapses in the central nervous system, and then along a motor neuron to an effector organ that carries out a response.

The nervous system as a whole is divided into the **central nervous system,** made up of the brain and the spinal cord, and the **peripheral nervous system,** made up of the cranial and spinal nerves, which connects all parts of the body with the central nervous system. The spinal cord carries impulses to and from the brain. It is also a center for simple reflex activities in which responses are coordinated within the cord. The brain and related structures are the subject of Chapter 9.

A functional subdivision of the nervous system is the **autonomic nervous system,** which controls unconscious activities. This system regulates the actions of glands, smooth muscle, and the heart muscle. The autonomic nervous system has two divisions, the **sympathetic nervous system** and the **parasympathetic nervous system,** which generally have opposite effects on a given organ.

II. Topics for Review

A. Divisions of the nervous system
   1. Central nervous system
   2. Peripheral nervous system
   3. Autonomic nervous system
B. Nervous tissue
   1. The nerve cell and its fibers
   2. The nerve impulse
   3. The synapse
   4. Nerves
   5. The reflex arc
C. Spinal cord
   1. Structure
   2. Function
   3. Simple reflexes
D. Spinal nerves
   1. Location
   2. Branches
E. Autonomic nervous system
   1. Divisions
      a. Sympathetic
      b. Parasympathetic
   2. Functions

III. Matching Exercises

Matching only within each group, write the answers in the spaces provided.

Group A

synapse	neuron	nerve impulse
dendrite	root	plexus
receptor		

1. The scientific name for a nerve cell   _____

2. A nerve cell fiber that carries impulses toward the cell body is called a(n) _____

3. The point at which impulses are transmitted from one nerve cell to another _____

4. A branch of a spinal nerve that attaches to the spinal cord _____

5. An electric charge that spreads along the membrane of a nerve cell _____

6. A network formed by the larger anterior branches of a spinal nerve _____

7. The structure that receives a stimulus _____

Group B

craniosacral   reflex   neurotransmitter
thoracolumbar   ganglion   neurilemma
nerve   mixed   afferent

1. Term for nerve fibers that carry impulses toward the spinal cord or brain _____

2. A collection of nerve cell bodies located outside the central nervous system _____

3. Term that describes the sympathetic portion of the autonomic nervous system, based on where it originates _____

4. A bundle of nerve cell fibers located outside the central nervous system _____

5. The sheath found around some nerve fibers that aids in regeneration if the fiber is damaged _____

6. Term that describes most nerves, notably the spinal nerves, because they contain both afferent and efferent fibers _____

7. Term that describes the parasympathetic portion of the autonomic nervous system, based on where it originates _____

8. A simple, automatic response that involves few neurons _____

9. A chemical that carries an impulse across a synapse _____

Group C

stretch reflex   brachial plexus   sciatic nerve
motor neurons   sensory fibers   cervical plexus
association neurons

1. The structures contained in the dorsal root of a spinal nerve _____

2. Another name for central neurons, those located inside the central nervous system _____

3. The type of response exemplified by the knee jerk _____

4. The largest branch of the lumbosacral plexus _____

5. The type of cells in the ventral gray horn of the spinal cord _____

6. The network of nerves that supplies the neck muscles _____

7. The network of nerves that supplies the upper extremities _____

Group D

parasympathetic nervous system   sympathetic chain   acetylcholine
sympathetic nervous system   adrenal   adrenergic
reflex arc

1. A cordlike strand of ganglia that extends along the spinal
   column _____

2. Adjective for a response activated by the neurotransmitter
   epinephrine _____

3. The neurotransmitter released from neurons of the parasympathetic nervous system _____

4. The system that promotes the fight-or-flight response _____

5. A gland that produces epinephrine _____

6. The system that stimulates the digestive and urinary tracts _____

7. A complete pathway through the nervous system from stimulus
   to response _____

IV. Multiple Choice

Select the best answer and write the letter of your choice in the blank.

1. A sudden electrical change in a nerve cell membrane that starts
   a nerve impulse is called a(n)   1. _____

   a. dendrite
   b. neurilemma
   c. action potential
   d. receptor
   e. effector

2. Which of the following substances is a neurotransmitter?   2. _____

   a. melanin
   b. acetylcholine

c. sebum

d. myelin

e. actin

3. Myelinated fibers are found in the

3. _____

    a. dorsal horn

    b. ventral horn

    c. central canal

    d. white matter

    e. gray matter

4. Cell bodies of sensory neurons are collected in ganglia located on each

4. _____

    a. dorsal root

    b. sympathetic chain

    c. ventral root

    d. effector organ

    e. plexus

5. Motor nerves are also described as

5. _____

    a. afferent

    b. sensory

    c. tracts

    d. ascending

    e. efferent

6. Which of the following are effectors of the nervous system?

6. _____

    a. sensory neurons and ganglia

    b. receptors and neurotransmitters

    c. synapses and dendrites

    d. adipose tissue and tendons

    e. muscles and glands

7. The voluntary nervous system controls

7. _____

    a. smooth muscle

    b. skeletal muscle

    c. glands

    d. cardiac muscle

    e. visceral muscle

V. Labeling

For each of the following illustrations, write the name or names of each labeled part on the numbered lines.

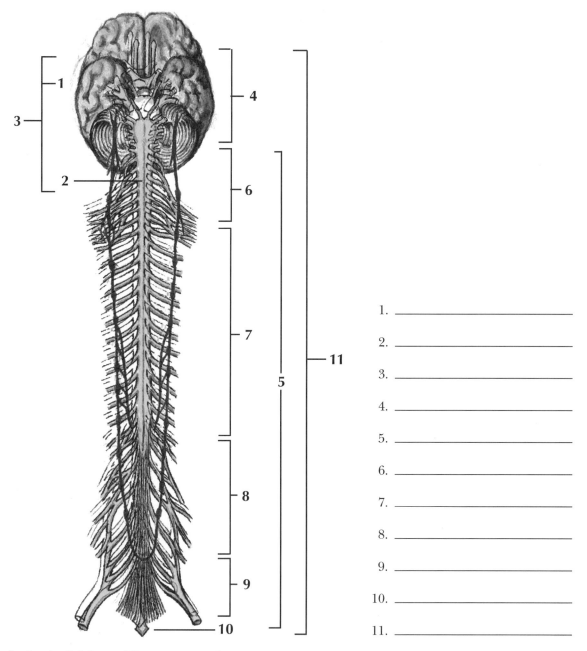

Anatomic divisions of the nervous system

1. _____

2. _____

3. _____

4. _____

5. _____

6. _____

7. _____

8. _____

9. _____

10. _____

11. _____

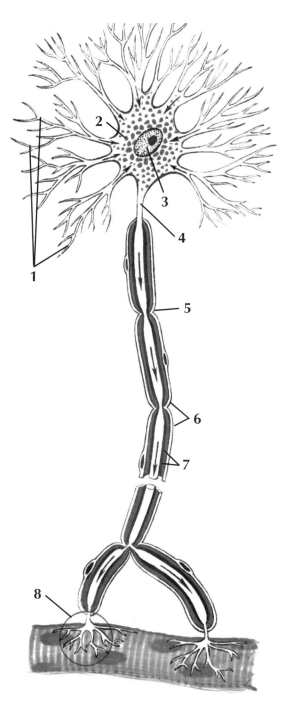

Diagram of a motor neuron

1. _____   5. _____

2. _____   6. _____

3. _____   7. _____

4. _____   8. _____

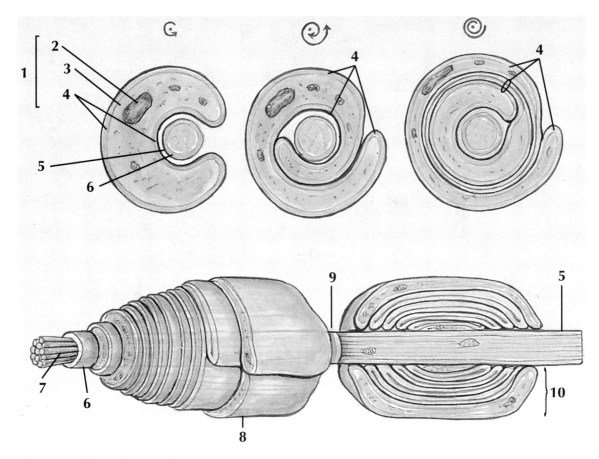

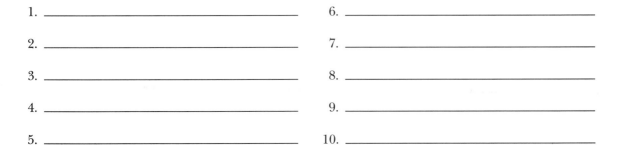

Formation of a myelin sheath

1. _____

2. _____

3. _____

4. _____

5. _____

6. _____

7. _____

8. _____

9. _____

10. _____

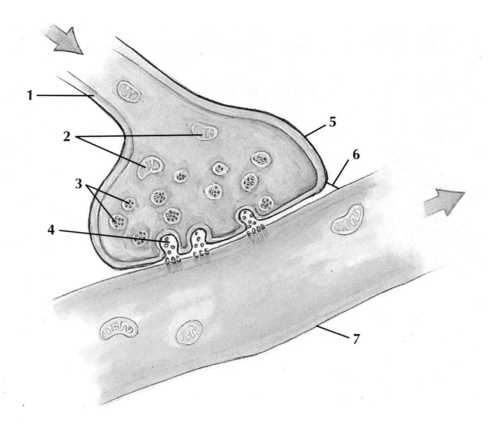

Close-up of a synapse

1. _____   5. _____

2. _____   6. _____

3. _____   7. _____

4. _____

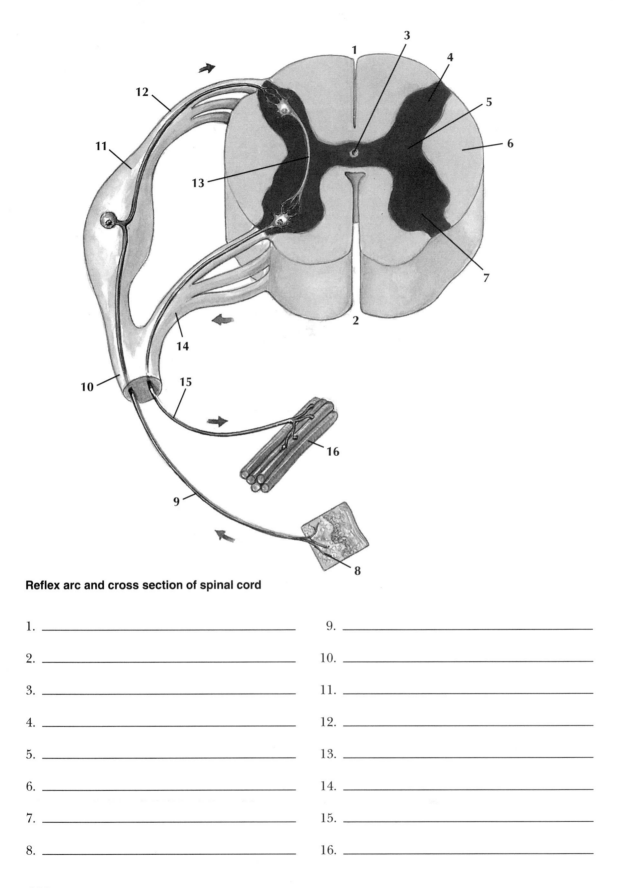

Reflex arc and cross section of spinal cord

1. _____

2. _____

3. _____

4. _____

5. _____

6. _____

7. _____

8. _____

9. _____

10. _____

11. _____

12. _____

13. _____

14. _____

15. _____

16. _____

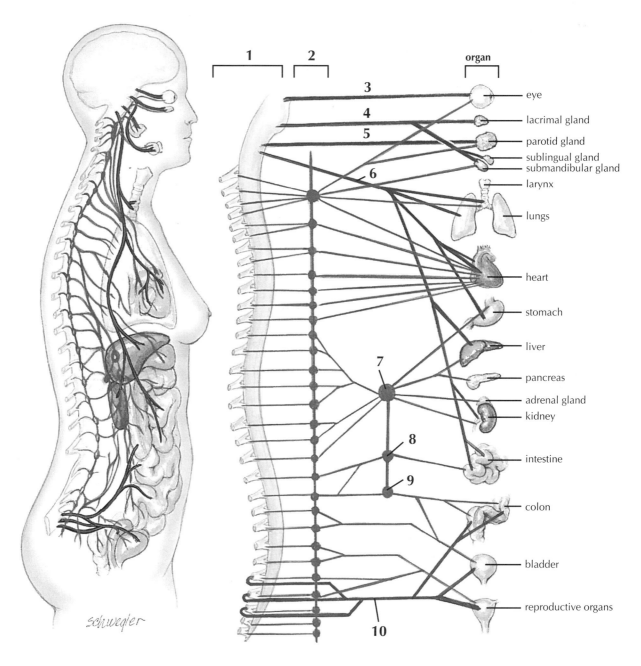

organ

3 — eye

4 — lacrimal gland

5 — parotid gland

6 — sublingual gland
submandibular gland

— larynx

— lungs

— heart

— stomach

— liver

7 — pancreas

— adrenal gland
— kidney

8 — intestine

9 — colon

— bladder

10 — reproductive organs

schwegler

Autonomic nervous system

1. _____   6. _____

2. _____   7. _____

3. _____   8. _____

4. _____   9. _____

5. _____   10. _____

VI. True-False

For each question, write T for true and F for false in the blank to the left of each number. If a statement is false, correct it by replacing the <u>underlined</u> term and write the correct statement in the blanks below the question.

_____ 1. Sensory nerves are also described as <u>afferent</u>.

_____

_____

_____ 2. The cranial and spinal nerves make up the <u>peripheral</u> nervous system.

_____

_____

_____ 3. The <u>somatic</u> nervous system is also called the voluntary nervous system.

_____

_____

_____ 4. Myelinated fibers form the <u>gray</u> matter of the nervous system.

_____

_____

_____ 5. A <u>sensory</u> neuron carries impulses away from the central nervous system.

_____

_____

_____ 6. A <u>tract</u> is a bundle of neuron fibers within the central nervous system.

_____

_____

_____ 7. An <u>axon</u> carries nerve impulses toward the cell body.

_____

_____

_____ 8. Sensory impulses enter the <u>dorsal</u> horn of the spinal cord.

_____

_____

_____ 9. The <u>parasympathetic</u> nervous system arises from the thoracic and lumbar regions of the spinal cord.

_____

_____

VII. Completion Exercise

Write the word or phrase that correctly completes each sentence.

1. The brain and spinal cord together are referred to as the _____

2. The fat-like covering of some nerve fibers is called _____

3. A nerve cell is also called a(n) _____

4. The small channel in the center of the spinal cord that contains cerebrospinal fluid is the _____

5. A nerve fiber that conducts impulses away from the cell body is a(n) _____

6. A specialized nerve ending that can detect a stimulus is a(n) _____

7. Dilation of the bronchial tubes is increased by the part of the autonomic nervous system called the _____

8. The largest branch of the lumbosacral plexus is the _____

VIII. Practical Applications

Study each discussion. Then write the appropriate word or phrase in the space provided.

1. Miss B was being treated for lung cancer. Her symptoms included pain in the back and weakness of the right arm. X-ray studies showed a tumor encroaching on the space containing the spinal cord. This space is called the _____

2. Mr. C was involved in a serious automobile accident. A study of shoulder x-rays showed a fractured humerus that cut into the network of nerves that supplies the upper extremity. This mass of nerves is called the _____

3. Miss D fell down a staircase. A diagnostic study of her injuries included the removal of fluid from the space below the end of the spinal cord. The spinal cord ends in the region of the spine named the _____

4. John, age 16, was brought to the emergency room after experimenting with street drugs. His symptoms included a rapid heart rate, dilated pupils, high blood pressure, increased body temperature, and other indications of excessive stimulation of the special part of the nervous system that controls involuntary functions of the viscera. This system is called the _____

IX. Short Essays

1. Differentiate between the visceral nervous system and the somatic nervous system.

   _____

   _____

   _____

   _____

2. Describe a synapse between two neurons.

   _____

   _____

   _____

   _____

3. What is a reflex arc and what are its components?

   _____

   _____

   _____

   _____

Memmler, RL, Cohen, BJ, Wood, DL. *STUDY GUIDE FOR STRUCTURE AND FUNCTION OF THE HUMAN BODY*, 6/e, © 1996, Lippincott-Raven Publishers

CHAPTER 9

The Nervous System: The Brain and Cranial Nerves

I. Overview

The brain is the largest mass of nerve tissue in the body and contains billions of nerve cells. It weighs about 3 pounds and is very fragile. It consists of the two cerebral hemispheres, the diencephalon, the brain stem, and the cerebellum, each with specific functions.

The brain and spinal cord are covered by three layers of fibrous membranes called the *meninges.* Aiding in the protection of the brain and cord is the *cerebrospinal fluid,* which is produced by the choroid plexuses in four ventricles (spaces) within the brain.

Connected with the brain are 12 pairs of *cranial nerves*, most of which supply structures in the head. Some of these, like all the spinal nerves, are mixed nerves containing both sensory and motor fibers. Some of the cranial nerves are made entirely of sensory fibers, whereas others are motor in function. Even those that contain both types of fibers do not have the mixture of fibers found in spinal nerves.

II. Topics for Review

A. The meninges
B. Cerebrospinal fluid
C. Divisions of the brain
    1. Cerebral hemispheres
       a. Lobes
       b. Cortex

2. Diencephalon
    a. Thalamus
    b. Hypothalamus
3. Brain stem
    a. Midbrain
    b. Pons
    c. Medulla oblongata
4. Cerebellum
D. Cranial nerves

III. Matching Exercises

Matching only within each group, write the answers in the spaces provided.

Group A

gyri   meninges   hemisphere
ventricles   brain stem   lobes
thalamus   sulci

1. The term for each half of the cerebrum   _____

2. The part of the brain composed of the midbrain, pons,
   and medulla   _____

3. The elevated portions of the cerebral cortex   _____

4. The spaces within the brain where cerebrospinal fluid is
   produced   _____

5. The collective name for the three brain coverings   _____

6. The individual subdivisions of the cerebral hemispheres that
   regulate specific functions   _____

7. The region of the diencephalon that acts as a relay center for
   sensory stimuli   _____

8. The shallow grooves in the cortex of the cerebrum   _____

Group B

dura mater   arachnoid   pia mater
arachnoid villi   choroid plexus   subarachnoid space

1. The innermost layer of the meninges, the delicate membrane
   in which there are many blood vessels   _____

2. The weblike middle meningeal layer   _____

3. The outermost layer of the meninges, which is the thickest
   and toughest   _____

4. The vascular network in a ventricle, which forms cerebro-spinal fluid

_____

5. The area in which cerebrospinal fluid (CSF) collects before its return to the blood

_____

6. The projections in the dural sinuses through which CSF is returned to the blood

_____

Group C

cerebellum	occipital lobe	motor cortex
corpus callosum	parietal lobe	temporal lobe
medulla oblongata		

1. The area in each frontal lobe, near the central sulcus, that controls the voluntary muscles

_____

2. Location of a sensory area for interpretation of pain, touch, and temperature

_____

3. The portion of the cerebral cortex where auditory impulses are interpreted

_____

4. The portion of the cerebral cortex where visual impulses from the retina are interpreted

_____

5. The division of the brain that coordinates voluntary muscles and helps to maintain balance

_____

6. A band of white matter that acts as a bridge between the cerebral hemispheres

_____

7. The part of the brain between the pons and the spinal cord

_____

Group D

facial nerve	vagus nerve	vestibulocochlear nerve
optic nerve	hypoglossal nerve	oculomotor nerve
accessory nerve	olfactory nerve	trigeminal nerve

1. The nerve that controls tongue muscles

_____

2. The sensory nerve that carries visual impulses

_____

3. The nerve that controls contraction of most eye muscles

_____

4. The nerve with three branches that carries general sensory impulses from the face and head

_____

5. The nerve that supplies most of the organs in the thoracic and abdominal cavities

_____

6. The nerve that contains sensory fibers for hearing _____

7. The nerve that supplies the muscles of facial expression _____

8. The nerve that carries motor impulses to two neck muscles _____

9. The nerve that carries impulses for the sense of smell _____

Group E

diencephalon	central sulcus	lateral sulcus
internal capsule	limbic system	medulla oblongata

1. The groove that runs at right angles to the longitudinal fissure between the frontal and parietal lobes _____

2. The groove that curves along the side of each hemisphere and separates the temporal lobe from the frontal and parietal lobes _____

3. The portion of the brain that contains the thalamus and hypothalamus _____

4. A crowded strip of nerve fibers that carries messages to and from the brain cortex _____

5. The region consisting of portions of the cerebrum and diencephalon that is involved in emotional states and behavior _____

6. The location of the vasomotor center, which regulates smooth muscle contraction in blood vessels _____

IV. Multiple Choice

Select the best answer and write the letter of your choice in the blank.

1. The pia mater is

    a. the innermost layer of the meninges
    b. the middle layer of the meninges
    c. the network of vessels that produces cerebrospinal fluid
    d. a raised area on the surface of the cerebrum
    e. the part of the brain that connects with the spinal cord

2. _____

2. Which of the following is not a lobe of the cerebrum?

    a. parietal
    b. frontal
    c. cortical
    d. occipital
    e. temporal

3. _____

3. The reticular formation is

   a. a venous channel in the brain
   b. a region of the limbic system that controls wakefulness and sleep
   c. a deep groove that divides the cerebral hemispheres
   d. the part of the temporal lobe concerned with the sense of smell
   e. the fifth lobe of the cerebrum

4. _____

4. The forward part of the brain stem that contains relay centers for eye and ear reflexes is the

   a. cortex
   b. thalamus
   c. cerebellum
   d. midbrain
   e. cerebrum

5. _____

5. The glossopharyngeal nerve supplies the

   a. face and eye
   b. ear and pharynx
   c. ear and nose
   d. tongue and pharynx
   e. lower jaw and thoracic organs

6. _____

V. Labeling

For each of the following illustrations, write the name or names of each labeled part on the numbered lines.

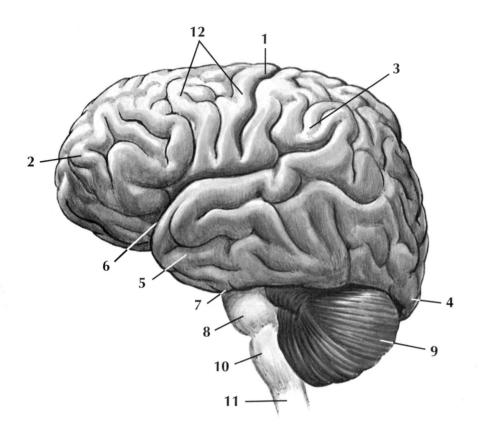

External surface of the brain

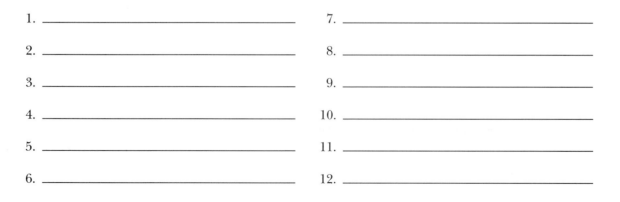

1. _____   7. _____

2. _____   8. _____

3. _____   9. _____

4. _____   10. _____

5. _____   11. _____

6. _____   12. _____

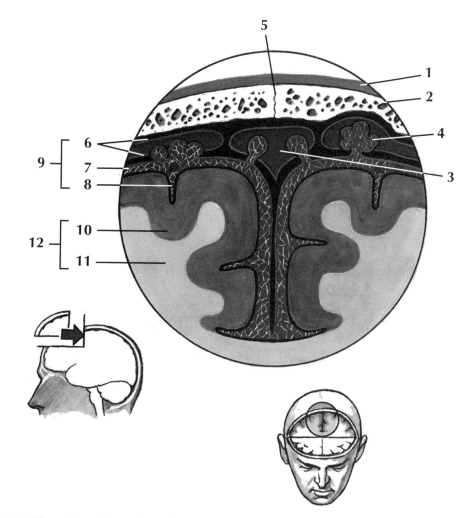

Frontal (coronal) section of top of head to show meninges and related parts

1. _____

2. _____

3. _____

4. _____

5. _____

6. _____

7. _____

8. _____

9. _____

10. _____

11. _____

12. _____

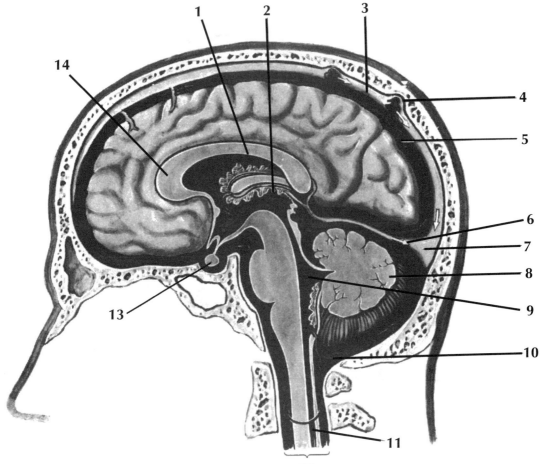

Flow of cerebrospinal fluid

1. _____
2. _____
3. _____
4. _____
5. _____
6. _____
7. _____

8. _____
9. _____
10. _____
11. _____
12. _____
13. _____
14. _____

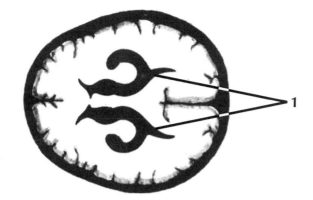

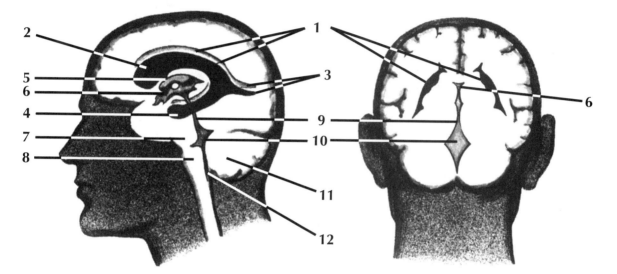

Ventricles of the brain

1. _____   7. _____

2. _____   8. _____

3. _____   9. _____

4. _____   10. _____

5. _____   11. _____

6. _____   12. _____

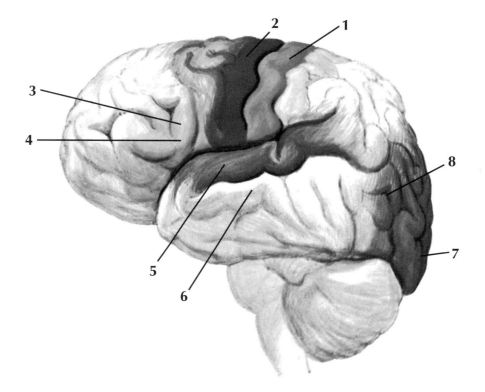

Functional areas of the cerebral cortex

1. _____   5. _____

2. _____   6. _____

3. _____   7. _____

4. _____   8. _____

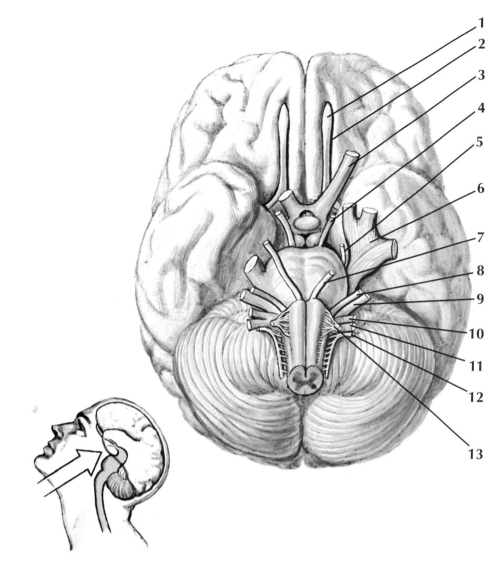

Base of brain, showing cranial nerves

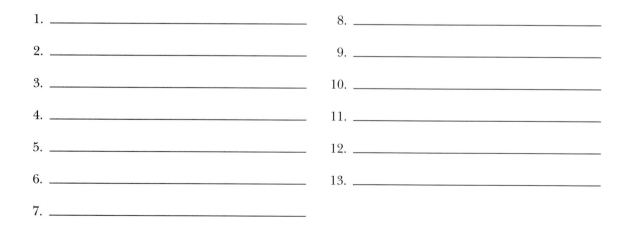

1. _____

2. _____

3. _____

4. _____

5. _____

6. _____

7. _____

8. _____

9. _____

10. _____

11. _____

12. _____

13. _____

VI. True-False

For each question, write T for true and F for false in the blank to the left of each number. If a statement is false, correct it by replacing the underlined term and write the correct statement in the blanks below the question.

_____ 1. The raised areas on the surface of the cerebrum are called <u>sulci</u>.

_____

_____

_____ 2. The <u>dura mater</u> is the innermost layer of the meninges.

_____

_____

_____ 3. The visual area of the cerebral cortex is in the <u>occipital</u> lobe.

_____

_____

_____ 4. The area outside the dura mater is described as <u>subdural</u>.

_____

_____

_____ 5. The <u>pons</u> is the middle portion of the brain stem.

_____

_____

_____ 6. The trigeminal nerve is the <u>fifth</u> (V) cranial nerve.

_____

_____

_____ 7. The vestibulocochlear nerve (VIII) is a <u>sensory</u> nerve.

_____

_____

VII. Completion Exercise

Write the word or phrase that correctly completes each sentence.

1. The largest part of the brain is the _____

2. The three layers of membranes that surround the brain and spinal cord are called the _____

3. The tough outermost membrane surrounding the brain is the _____

4. The clear liquid that helps to support and protect the brain and spinal cord is _____

5. The number of pairs of cranial nerves is _____

6. The spinal cord connects with the part of the brain called the _____

7. The thin layer of gray matter that forms the surface of each cerebral hemisphere is the _____

8. The four chambers within the brain where cerebrospinal fluid is produced are called _____

9. The region of the diencephalon that helps maintain homeostasis (*e.g.*, water balance, appetite, and body temperature) and controls the autonomic nervous system is the _____

10. Records of the electrical activity of the brain can be made with an instrument called a(n) _____

VIII. Practical Applications

Study each discussion. Then write the appropriate word or phrase in the space provided.

1. Mrs. N, age 67, had been found lying on the floor, unconscious. She had a history of high blood pressure and was diagnosed as having had a stroke. Now she is unable to speak or write. The motor areas for communication by speech and writing are located in the cortex of the _____

2. Mr. H, age 42, had been suffering from severe headaches for 8 weeks. A brain study involving three-dimensional x-rays was ordered. Such a study is called _____

3. Young A, age 10, was brought to the emergency room after he had fallen from his bicycle and hit his head so that there was external bleeding. Studies were done to determine the possibility of internal damage due to a tear in the wall of a venous channel in the dura mater. Such a channel is called a(n) _____

4. Miss K, age 16, told her physician that she had difficulty maintaining coordination and balance. These symptoms might indicate there was a tumor in the dorsal subdivision of the brain known as the _____

5. Ms. J, age 32, complained of embarrassing distortion of her face due to a paralysis of facial muscles on one side. There was distressing speech difficulty and an excessive amount of tears flowing onto her cheeks. A diagnosis of Bell's palsy was made. This paralysis involves the VIIth cranial nerve named the _____

IX. Short Essays

1. Describe the structures that protect the brain and spinal cord.

_____

_____

_____

_____

2. List some areas of the cerebral cortex that are involved in communication.

_____

_____

_____

_____

3. List some functions of the structures in the diencephalon.

_____

_____

_____

_____

Memmler, RL, Cohen, BJ, Wood, DL. *STUDY GUIDE FOR STRUCTURE AND FUNCTION OF THE HUMAN BODY*, 6/e, © 1996, Lippincott-Raven Publishers

CHAPTER 10

The Sensory System

I. Overview

Through the functioning of the *sensory receptors,* we are made aware of changes taking place both internally (within the body) and externally (outside the body). Any change that produces a response in the nervous system is termed a *stimulus.*

The *special senses,* so-called because the receptors are limited to a few specialized sense organs, include the senses of vision, hearing, equilibrium, taste, and smell. The receptors of the eye are the *rods and cones* located in the retina. The hearing receptors are found in a portion of the inner ear called the *cochlea.* Receptors for the chemical senses of taste and smell are located on the tongue and in the upper part of the nose respectively.

The *general senses* are scattered throughout the body; they respond to pressure, temperature, pain, touch, and position. Receptors for the sense of position, known as *proprioceptors,* are found in muscles, tendons, and joints.

The nerve impulses generated in a receptor cell by a stimulus must be carried to the central nervous system by way of a sensory (afferent) neuron. Here the information is processed and a suitable response is made.

II. Topics for Review

A. The eye
   1. Protective structures of eyeball
   2. Coats of eyeball; sclera, choroid, retina
   3. Pathway of light rays
   4. Muscles of the eye
      a. Extrinsic
      b. Intrinsic
   5. Nerve supply to the eye

B. The ear
   1. Sections
      a. External ear: pinna, auditory canal, tympanic membrane
      b. Middle ear: ossicles
      c. Internal ear: vestibule, semicircular canals, cochlea
   2. Receptors
      a. Hearing
      b. Equilibrium
C. Other organs of special sense
   1. Taste receptors
   2. Smell receptors
D. General senses
   1. Pressure
   2. Temperature
   3. Touch
   4. Pain
   5. Position

III. Matching Exercises

Matching only within each group, write the answers in the spaces provided.

Group A

cornea	aqueous humor	choroid
accommodation	rods	cones
retina	vitreous body	

1. The innermost coat of the eyeball, the nervous tissue layer that includes the receptors for the sense of vision   _____

2. The vascular, pigmented, middle tunic of the eyeball   _____

3. The part of the eye that light rays pass through first as they enter the eye   _____

4. The watery fluid that fills much of the eyeball in front of the crystalline lens   _____

5. The jellylike material located behind the crystalline lens that maintains the spherical shape of the eyeball   _____

6. The vision receptors that function in dim light   _____

7. The vision receptors that are sensitive to color   _____

8. The process by which the lens becomes thicker to bend light rays for near vision   _____

Group B

iris	pupil	ciliary body
media	sclera	receptor
optic disk	conjunctiva	

1. The opaque outermost layer of the eyeball made of firm, tough connective tissue

_____

2. The central opening in the iris

_____

3. All the transparent refracting parts of the eye

_____

4. A part of the nervous system that detects a stimulus

_____

5. The membrane that lines the eyelids

_____

6. Another name for the blind spot, the region where the optic nerve connects with the eye

_____

7. The muscle that alters the shape of the lens for accommodation

_____

8. The colored part of the eye that regulates the size of the pupil

_____

Group C

fovea centralis   lacrimal gland   intrinsic
sphincter   refraction   crystalline lens
extrinsic

1. Term for the muscles located outside the eyeball that are attached to bones of the orbit and to the sclera

_____

2. A circular muscle, such as the muscle of the iris

_____

3. The bending of light rays so that light from a large area can be focused on a small surface

_____

4. The structure that produces tears

_____

5. The depressed area in the retina that is the point of clearest vision

_____

6. The part of the eye that is adjusted to focus light for near and far vision

_____

7. Term that describes the muscles of the iris and ciliary body because they are located entirely within the eyeball

_____

Group D

oval window   endolymph   eustachian tube
ossicles   pinna   tympanic membrane
perilymph

1. The scientific name for the eardrum

_____

2. The three small bones within the middle ear cavity

_____

3. The membrane-covered structure that conducts sound waves from the stapes to the fluid of the internal ear _____

4. The passageway that connects the middle ear cavity with the throat _____

5. The fluid of the inner ear contained within the bony labyrinth and surrounding the membranous labyrinth _____

6. The fluid contained within the membranous labyrinth _____

7. Another name for the projecting part, or auricle, of the ear _____

Group E

optic nerve   vestibule   ophthalmic nerve
oculomotor nerve   cochlear duct   cochlear nerve
equilibrium

1. The location of the organ of hearing _____

2. The branch of the vestibulocochlear nerve that carries hearing impulses _____

3. The entrance area that communicates with the cochlea and that is next to the oval window _____

4. The nerve that carries visual impulses from the retina to the brain _____

5. The branch of the fifth cranial nerve that carries impulses of pain, touch, and temperature from the eye to the brain _____

6. The largest of the three cranial nerves that carry motor fibers to the eyeball muscles _____

7. The sense that is located in the semicircular canals and the vestibule _____

Group F

glossopharyngeal nerve   ceruminous   olfactory nerve
adaptation   night blindness   tactile corpuscles
proprioceptors

1. One of the two nerves involved in the sense of taste _____

2. A condition that may result from a deficiency of vitamin A _____

3. Receptors for the sense of touch _____

4. An adjustment to the environment so that one does not feel a sensation so acutely if a stimulus is continued   _____

5. Term that describes the wax glands located in the external auditory canal   _____

6. The first cranial nerve, which conducts impulses from smell receptors   _____

7. Receptors that transmit information on the position of body parts   _____

IV. Multiple Choice

Select the best answer and write the letter of your choice in the blank.

1. The term *lacrimation* refers to the secretion of   1. _____

   a. mucus
   b. wax
   c. vitreous humor
   d. tears
   e. aqueous humor

2. The nerves involved in the sense of taste are the   2. _____

   a. vagus and vestibulocochlear
   b. oculomotor and trigeminal
   c. trochlear and olfactory
   d. ocular and abducens
   e. facial and glossopharyngeal

3. A person who has a lack of cones in the retina will suffer from   3. _____

   a. blindness
   b. color blindness
   c. presbyopia
   d. glaucoma
   e. trachoma

4. The organ of Corti is the receptor for   4. _____

   a. taste
   b. smell
   c. hearing
   d. pressure
   e. equilibrium

V. Labeling

For each of the following illustrations, write the name or names of each labeled part on the numbered lines.

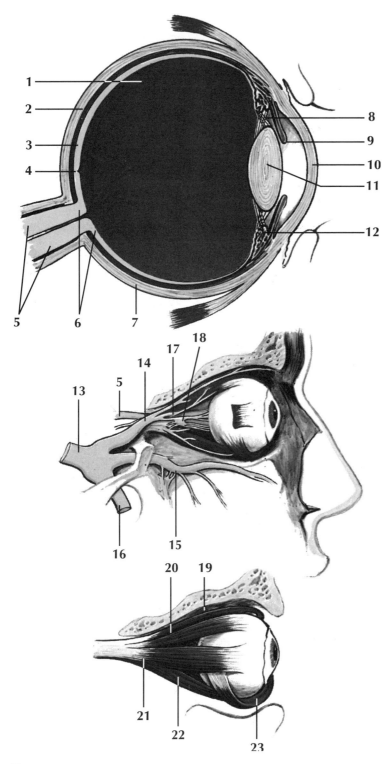

The eye

1. _____
2. _____
3. _____
4. _____
5. _____
6. _____
7. _____
8. _____
9. _____
10. _____
11. _____
12. _____
13. _____
14. _____
15. _____
16. _____
17. _____
18. _____
19. _____
20. _____
21. _____
22. _____
23. _____

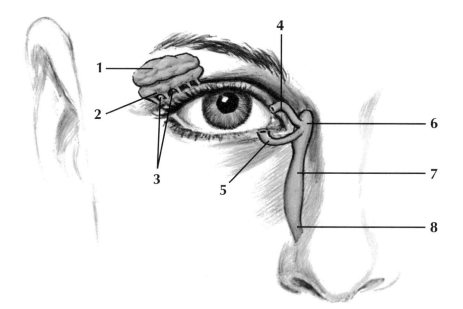

Lacrimal apparatus

1. _____   5. _____

2. _____   6. _____

3. _____   7. _____

4. _____   8. _____

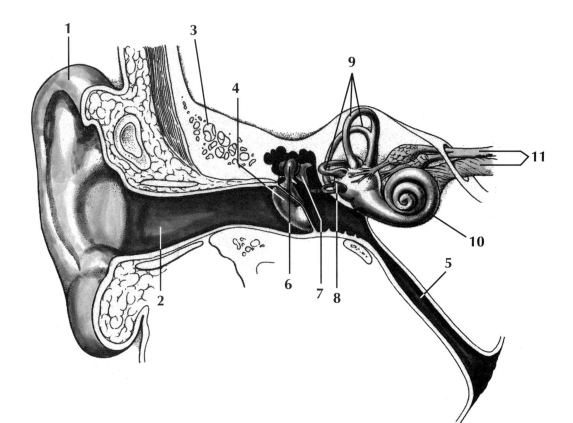

The ear

1. _____

2. _____

3. _____

4. _____

5. _____

6. _____

7. _____

8 _____

9. _____

10. _____

11. _____

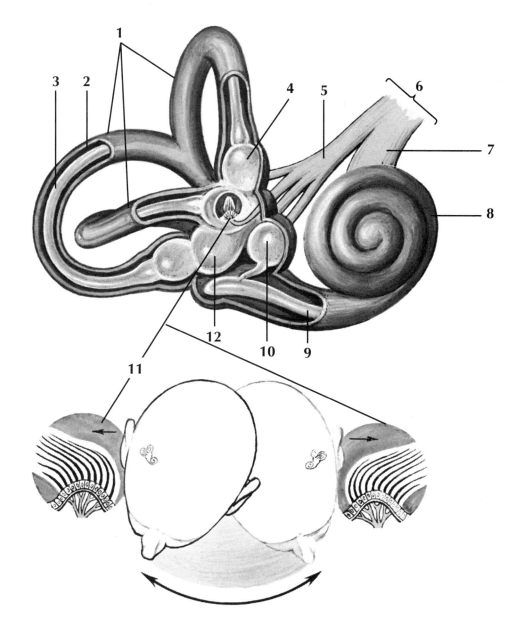

The internal ear

1. _____

2. _____

3. _____

4. _____

5. _____

6. _____

7. _____

8. _____

9. _____

10. _____

11. _____

12. _____

VI. True-False

For each question, write T for true and F for false in the blank to the left of each number. If a statement is false, correct it by replacing the underlined term and write the correct statement in the blanks below the question.

_____ 1. The middle coat, or tunic, of the eye is the <u>sclera</u>.

_____

_____

_____ 2. The are <u>six</u> extrinsic muscles connected to each eye.

_____

_____

_____ 3. The iris is an <u>intrinsic</u> muscle of the eye.

_____

_____

_____ 4. The <u>rods</u> of the eye function in bright light and detect color.

_____

_____

_____ 5. When the eyes are exposed to a bright light, the pupils <u>contract</u>.

_____

_____

_____ 6. The <u>stapes</u> is in contact with the oval window of the ear.

_____

_____

_____ 7. <u>Perilymph</u> is the fluid that fills the membranous labyrinth of the ear.

_____

_____

_____ 8. The <u>aqueous</u> humor is the fluid in front of the lens of the eye

_____

_____

VII. Completion Exercise

Write the word or phrase that correctly completes each sentence.

1. The nerve fibers of the vestibular and cochlear nerves join to
   form the nerve called the _____

2. The inner ear spaces contain fluids involved in the transmission
   of sound waves. The one that is inside the membranous cochlea
   and that stimulates the receptors is the _____

3. The very widely distributed free nerve endings are the receptors
   for the most important protective sense, the sense of _____

4. The tactile corpuscles are the receptors for the sense of _____

5. The gustatory sense is the sense of _____

6. The nerve endings that aid in judging position and changes in
   location of body parts are the _____

7. The sense of position is partially governed by several structures
   in the internal ear, including two small sacs in the vestibule and
   the three _____

8. When you enter a darkened room, it takes a while for the rods
   to begin to function. This interval is known as the period of _____

VIII. Practical Applications

Study each discussion. Then write the appropriate word or phrase in the space
provided.

Group A

While observing patients in the emergency ward, the student nurse noted the
following cases.

1. Ten-year-old K had been riding his bicycle while he threw glass
   bottles to the sidewalk. A fragment of glass flew into one eye.
   Examination at the hospital showed that there was a cut in the
   transparent window of the eye, the _____

2. On further examination of K, the colored part of the eye was
   seen to protrude from the wound. This part is the _____

3. K's treatment included antiseptics, anesthetics, and suturing of
   the wound. Medication was instilled in the saclike structure at
   the front of the eyeball. This sac is lined with a thin epithelial
   membrane, the _____

4. A construction worker, Mr. J, was admitted because of an accident in which a piece of steel penetrated his eyeball and caused such an extensive wound that material from the inside of the eyeball oozed out. Mr. J tried to relieve the pain by forcing the jellylike material out through the wound at the front and side of his eyeball. This matter, which maintains the shape of the eyeball, is called the _____

Group B

An ear, nose, and throat specialist treated the following patients one morning.

1. Mrs. B complained of some hearing loss and a sense of fullness in her outer ear. Examination revealed that her ear canal was plugged with hardened ear wax, which is scientifically called _____

2. Mr. J, age 72, complained of gradually increasing hearing loss, although he had no symptoms of pain or other problems related to the ears. Examination revealed that his hearing loss was due to nerve damage. The cranial nerve that carries hearing impulses to the brain is called the _____

3. Baby L was brought in by his mother because he awakened crying and holding the right side of his head. He had been suffering from a cold, but now he seemed to be in pain. Examination revealed a bulging red eardrum. The eardrum is also called the _____

4. Elderly Mr. N had a hearing loss (presbycusis) due to atrophy of the nerve endings located in the spiral-shaped part of the internal ear, a part of the ear that is known as the _____

IX. Short Essays

1. Compare the general and special senses and give examples of each.

_____

_____

_____

_____

2. Describe the structures that protect the eye.

_____

_____

_____

_____

3. Describe some changes that occur in the eye with age.

_____

_____

_____

_____

Memmler, RL, Cohen, BJ, Wood, DL. *STUDY GUIDE FOR STRUCTURE AND FUNCTION OF THE HUMAN BODY*, 6/e, © 1996, Lippincott-Raven Publishers

CHAPTER 11

The Endocrine System: Glands and Hormones

I. Overview

Hormones are chemical messengers that have specific regulatory effects on certain other cells or organs in the body, the *target tissue*. Although hormones are produced by many tissues, for certain glands hormone secretion is the primary function. These are the *endocrine glands* or ductless glands, including the pituitary (hypophysis), thyroid, parathyroids, adrenals, pancreas, and gonads. Together these comprise the *endocrine system*.

The endocrine system and the nervous system are the main coordinating and controlling systems of the body. Both are activated, for example, in helping the body respond to stress. These two systems meet in the *hypothalamus,* a region of the diencephalon of the brain. The hypothalamus, which is directly above and connected to the pituitary, governs both lobes of this gland. Hormones released from the pituitary, in turn, stimulate other endocrine glands. The main mechanism for controlling hormone secretion is *negative feedback,* in which hormone levels, or substances released as a result of hormone action, serve to regulate the production of that hormone.

Other structures that secrete hormones include the thymus, pineal, kidney, stomach, small intestine, and heart. *Prostaglandins* are hormones produced by cells throughout the body. They have a variety of effects and are currently under study.

II. Topics for Review

A. General characteristics of the endocrine system
B. Hormones
    1. Function
    2. Chemical makeup
    3. Regulation
C. The endocrine glands and their hormones
    1. Control of the pituitary by the hypothalamus
D. Other hormone-producing organs
E. Hormones and stress

III. Matching Exercises

Matching only within each group, write the answers in the spaces provided.

Group A

parathyroid glands   thyroid   islets of Langerhans
hormone   medulla   suprarenal glands

1. A substance produced by an endocrine gland   _____

2. Another name for the adrenal glands   _____

3. The groups of hormone-secreting cells scattered throughout the pancreas   _____

4. The largest of the endocrine glands, located in the neck   _____

5. The inner part of the adrenal gland   _____

6. The tiny glands located behind the thyroid gland   _____

Group B

thyroxine   negative feedback   pineal body
hypothalamus   calcitonin   target tissue
amino acids   adrenaline

1. The gland in the brain that is regulated by light   _____

2. The part of the brain that controls the pituitary gland   _____

3. The building blocks of protein hormones   _____

4. The specific cells on which a hormone works   _____

5. The hormone that is the main regulator of heat and energy production in the body   _____

6. The hormone produced by the thyroid gland that is active in calcium metabolism   _____

7. The type of self-controlling mechanism that regulates hormone production

    _____

8. The common name for epinephrine

    _____

Group C

iodine	insulin	parathyroid hormone
adrenal	releasing hormone	pituitary

1. The endocrine gland that is divided into anterior and posterior lobes

    _____

2. A secretion that raises the level of calcium in the blood

    _____

3. A hormone that lowers the level of sugar in the blood

    _____

4. The endocrine gland composed of an external cortex and an internal medulla, each with specific functions

    _____

5. The chemical element that is needed for the manufacture of thyroxine

    _____

6. A secretion from the hypothalamus that stimulates activity of the anterior lobe of the pituitary

    _____

Group D

steroids	placenta	anterior lobe
thymosin	oxytocin	glucagon
cortisol	calcium	

1. The part of the pituitary connected to the hypothalamus by a portal system

    _____

2. An organ present only during pregnancy that secretes hormones needed for normal development of the embryo

    _____

3. The chemical category that includes the sex hormones and the hormones of the adrenal cortex

    _____

4. The element regulated by hydroxycholecalciferol, a hormone produced from vitamin D

    _____

5. A hormone released by the adrenal cortex during stressful situations that acts to reduce inflammation

    _____

6. The hormone produced by the islets of Langerhans that raises blood sugar levels

    _____

7. The hormone that aids in maturation of the T cells needed for immunity

    _____

8. The hormone from the posterior pituitary that causes uterine contraction

_____

Group E

aldosterone	antidiuretic hormone	ACTH
luteinizing hormone	estrogen	somatotropin
epinephrine	kidney	

1. The main hormone of the adrenal medulla that, among other actions, raises blood pressure and increases the heart rate

_____

2. The hormone from the adrenal cortex that regulates the reabsorption of sodium and secretion of potassium in the kidney tubules

_____

3. A female sex hormone that most nearly parallels male testosterone in its action

_____

4. The hormone produced in the posterior lobe of the pituitary that regulates water reabsorption by the kidney

_____

5. An alternate name for growth hormone

_____

6. A gonadotropic hormone

_____

7. The anterior pituitary hormone that stimulates the adrenal cortex

_____

8. The organ that produces erythropoietin, a hormone that stimulates production of red blood cells

_____

IV. Multiple Choice

Select the best answer and write the letter of your choice in the blank.

1. Another name for the pituitary gland is

1. _____

    a. hypothalamus
    b. hypophysis
    c. thalamus
    d. cortex
    e. pineal

2. Gonadotropins act on the

2. _____

    a. kidneys and adrenal glands
    b. ovaries and testes
    c. hypothalamus and pituitary gland

d. pineal body and thymus gland

e. mammary glands and thyroid

3. Which of the following hormones is <u>not</u> produced by the anterior pituitary?

3. _____

a. prolactin
b. thyrotropin
c. oxytocin
d. luteinizing hormone
e. growth hormone

4. An androgen is a(n)

4. _____

a. female sex hormone
b. glucocorticoid
c. lymphocyte
d. male sex hormone
e. atrial hormone

5. The gland active in immunity is the

5. _____

a. thyroid
b. thymus
c. pineal
d. kidney
e. posterior pituitary

V. Labeling

For the following illustration, write the name or names of each labeled part on the numbered lines.

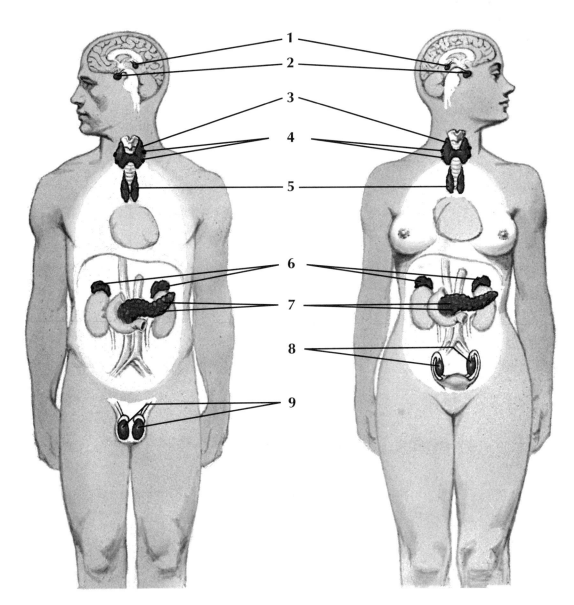

Main hormone-secreting organs

1. _____   6. _____

2. _____   7. _____

3. _____   8. _____

4. _____   9. _____

5. _____

VI. True-False

For each question, write T for true and F for false in the blank to the left of each number. If a statement is false, correct it by replacing the underlined term and write the correct statement in the blanks below the question.

_____ 1. Islet cells are found in the adrenal gland.

_____

_____

_____ 2. The ovaries and testes produce steroid hormones.

_____

_____

_____ 3. ADH and oxytocin are secreted by the anterior lobe of the pituitary.

_____

_____

_____ 4. Glucagon is the pancreatic hormone that lowers blood sugar levels.

_____

_____

_____ 5. Parathyroid hormone raises calcium levels in the blood.

_____

_____

_____ 6. Atrial natriuretic peptide (ANP) is produced by the kidneys.

_____

_____

VII. Completion Exercise

Write the word or phrase that correctly completes each sentence.

1. The region of the brain that controls the pituitary gland is the _____

2. The hormone epinephrine is produced by the _____

3. The hypothalamus stimulates the anterior pituitary to produce ACTH, which in turn stimulates hormone production by the _____

4. Local hormones that have a variety of effects, including the promotion of inflammation and the production of uterine contractions, are the

_____

5. When the level of glucose in the blood decreases to less than average, the islet cells of the pancreas release less insulin. The result is an increase in blood glucose. This is an example of the mechanism called

_____

VIII. Practical Applications

Study each discussion. Then write the appropriate word or phrase in the space provided.

1. Mr. J, age 23, required evaluation of pituitary function. As part of this evaluation, an x-ray examination was planned because of the possibility that a tumor was the cause of his excessive height of 7 feet as well as his abnormal weakness. The tests revealed that a pituitary tumor had resulted in the excess production of

_____

2. Seventeen-year-old Ms. K had never had a menstrual period. The cause was diagnosed as a deficiency of the ovarian hormones called

_____

3. Mrs. C, age 56, had been brought to the hospital in a coma, that is, she was unconscious and could not be aroused. Tests revealed that her blood sugar was abnormally high. Mrs. C's illness was diabetes mellitus, which is due to lack of the hormone

_____

4. Mrs. K consulted her doctor with complaints of weakness, weight loss, and chronic diarrhea. She had also noticed a bronze coloration of her skin. Laboratory tests showed abnormal blood electrolytes due to a deficiency of aldosterone from the

_____

5. Mr. J, age 40 was admitted to the hospital with the following symptoms: rapid heart rate, difficult breathing, nervousness, and heat intolerance. An enlargement of the neck indicated overactivity of the

_____

6. After surgery for his endocrine problem, Mr. J had contractions of the muscles of the hands and face. This was caused by the incidental surgical removal of the glands that control the release of calcium into the blood. The glands that maintain adequate blood calcium levels are the

_____

7. Mr. G had been taking high doses of steroid drugs for asthma for several years. He noted weight gain, bruising of skin, and muscle weakness. A routine urinalysis showed sugar in the urine. These symptoms indicated excessive amounts of the main glucocorticoid that is normally produced by the adrenal cortex and is named

_____

IX. Short Essays

1. Describe the characteristics of hormones.

   _____

   _____

   _____

   _____

2. Compare the anterior and the posterior lobes of the pituitary.

   _____

   _____

   _____

   _____

3. Tropic hormones, released by certain endocrine glands, stimulate other endocrine glands. Name several such tropic hormones and explain what each does.

   _____

   _____

   _____

   _____

4. Explain why hormones, although they circulate throughout the body, exercise their effects only on specific target cells.

   _____

   _____

   _____

   _____

IV

UNIT

Circulation and Body Defense

12. THE BLOOD
13. THE HEART AND HEART DISEASE
14. BLOOD VESSELS AND BLOOD CIRCULATION
15. THE LYMPHATIC SYSTEM AND IMMUNITY

Memmler, RL, Cohen, BJ, Wood, DL. *STUDY GUIDE FOR STRUCTURE AND FUNCTION OF THE HUMAN BODY*, 6/e, © 1996, Lippincott-Raven Publishers

The Blood

I. Overview

The blood maintains the internal environment in a constant state through its functions of transportation, regulation, and protection. Blood is composed of two elements: the liquid element, or *plasma*, and the *formed elements*, consisting of the cells and cellular products. The plasma is 90% water and 10% proteins, carbohydrates, lipids, and electrolytes. The formed elements are composed of the *erythrocytes*, which carry oxygen to the tissues by means of hemoglobin; the *leukocytes*, which defend the body against invaders; and the *platelets*, which are involved in the process of blood coagulation or clotting. The forerunners of the blood cells are called *stem cells*. These are formed in the red bone marrow, where they then develop into the various types of blood cells.

Blood *coagulation* is a protective mechanism that prevents blood loss when a blood vessel is ruptured by an injury. The first steps in the prevention of blood loss (hemostasis) include constriction of the blood vessels and formation of a platelet plug.

Should the quantity of blood in the body be severely reduced because of hemorrhage or disease, the cells suffer from lack of oxygen and nutrients. In such instances, a *transfusion* may be given after typing and matching the blood of the recipient and donor. (Red cells with different surface *antigens* [proteins] than the recipient's red cells will react with *antibodies* in the recipient's blood, causing harmful agglutination and destruction of the donated red cells.)

The presence or absence of *Rh factor*, a red blood cell protein, is also important in transfusions. If blood containing the Rh factor (Rh-positive) is given to a person whose blood lacks that factor (Rh-negative), the recipient may become sensitized to the protein; his blood cells will produce antibodies to counteract the foreign substance. If an Rh-negative mother becomes sensitized by an Rh-positive fetus, her antibodies may damage the red cells of the fetus in a later pregnancy unless proper treatment is given.

Numerous *blood studies* have been devised to measure the composition of blood. These include the hematocrit, tests for the amount of hemoglobin, cell counts, and coagulation studies. Modern laboratories are equipped with automatic counters, which rapidly and accurately count blood cells, and with automatic analyzers, which measure enzymes, electrolytes, and other constituents of blood serum.

II. Topics for Review

A. Functions of blood
B. Blood plasma and its functions
C. The formed elements and their functions
    1. Erythrocytes
        a. Structure
        b. Function
    2. Leukocytes
        a. Types
        b. Functions
    3. Platelets (thrombocytes)
D. Origin of blood cells
E. Hemostasis; blood clotting
F. Blood typing and blood transfusions
    1. Blood groups
        a. ABO
        b. Rh
G. Blood studies

III. Matching Exercises

Matching only within each group, write the answers in the spaces provided.

Group A

oxygen   plasma   carbon dioxide
hemoglobin   leukocyte   erythrocyte
thrombocyte

1. The liquid part of the blood   _____

2. A red blood cell   _____

3. A white blood cell   _____

4. Another name for a platelet, an element active in blood clotting   _____

5. The important gas that is transported by the blood from the lungs to all parts of the body   _____

6. The gaseous waste product carried by the blood to the lungs   _____

7. The pigment in red blood cells that carries oxygen   _____

Group B

albumin	neutrophils	antigens
hemostasis	Rh	red marrow

1. The connective tissue in bone that is the site of blood cell formation   _____

2. The most abundant protein in the blood   _____

3. The proteins on the surface of the red blood cells that limit the types of transfusions that can be given to a person   _____

4. The most numerous leukocytes in the blood   _____

5. The blood antigen involved in incompatibility between a mother and a fetus   _____

6. The name for the process that prevents blood loss   _____

Group C

serum	fibrinogen	hemolysis
megakaryocytes	hemoglobin	pus
agglutination	hyperglycemia	

1. The substance in red blood cells that contains iron   _____

2. The large cells that give off fragments known as platelets   _____

3. One of the plasma proteins that is activated as platelets disintegrate   _____

4. The process whereby cells become clumped when mixed with a specific antiserum   _____

5. The watery fluid that remains after a clot is removed   _____

6. The rupture of red blood cells   _____

7. A substance that often accumulates when leukocytes are actively destroying bacteria   _____

8. An excessive amount of sugar in the blood, as seen in unregulated diabetes   _____

Group D

hemocytometer	transfusion	4.5 to 5 million
centrifuge	electrolytes	hemorrhage
5,000 to 10,000	hematocrit	

1. Another term for profuse abnormal bleeding   ————————————

2. The transfer of blood or a blood component from one person to another   ————————————

3. A machine that is used to separate the blood cells from blood plasma   ————————————

4. The term for the salts dissolved in body fluids, such as plasma   ————————————

5. An apparatus used for manual counts of blood cells   ————————————

6. The average number of red blood cells per cubic millimeter   ————————————

7. The average number of white blood cells per cubic millimeter   ————————————

8. The test that measures the volume percentage of red blood cells in whole blood   ————————————

IV. Multiple Choice

Select the best answer and write the letter of your choice in the blank.

1. Carbon monoxide can block the ability of the blood to carry   1. ————————

    a. carbon dioxide
    b. oxygen
    c. hemoglobin
    d. albumin
    e. plasma

2. The proportion of red blood cells to white blood cells is about   2. ————————

    a. 100 to 1
    b. 700 to 1
    c. 6 to 1
    d. 2 to 1
    e. 5000 to 1

3. Which of the following is not a function of blood?   3. ————————

    a. transportation of waste products
    b. transportation of nutrients
    c. distribution of heat
    d. hemostasis
    e. manufacture of hormones

4. Which of the following is <u>not</u> a type of white blood cell?

    a. platelet
    b. eosinophil
    c. neutrophil
    d. monocyte
    e. lymphocyte

4. _____

5. Blood clotting occurs in a complex series of steps. The substance that finally forms the clot is

    a. anticoagulant
    b. thromboplastin
    c. antibody
    d. fibrin
    e. albumin

5. _____

6. Which of the following might result in an Rh incompatibility problem?

    a. an Rh-positive mother and an Rh-negative fetus
    b. an Rh-negative mother and an Rh-positive fetus
    c. an Rh-negative mother and a type AB fetus
    d. an Rh-positive mother and an Rh-negative father
    e. an Rh-positive mother and an Rh-positive father

6. _____

V. Labeling

For each of the following illustrations, write the name or names of each labeled part on the numbered lines.

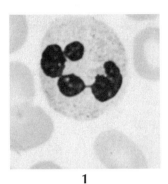

1

granules stain
lavender

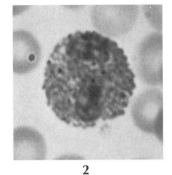

2

granules stain
bright pink

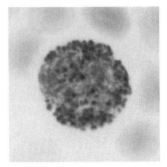

3

granules stain
dark blue

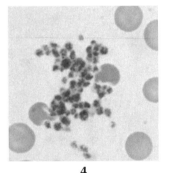

4

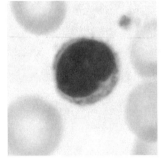

5

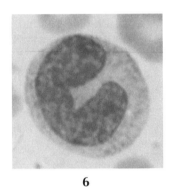

6

Blood cells

1. _____   4. _____

2. _____   5. _____

3. _____   6. _____

anti-A serum anti-B serum

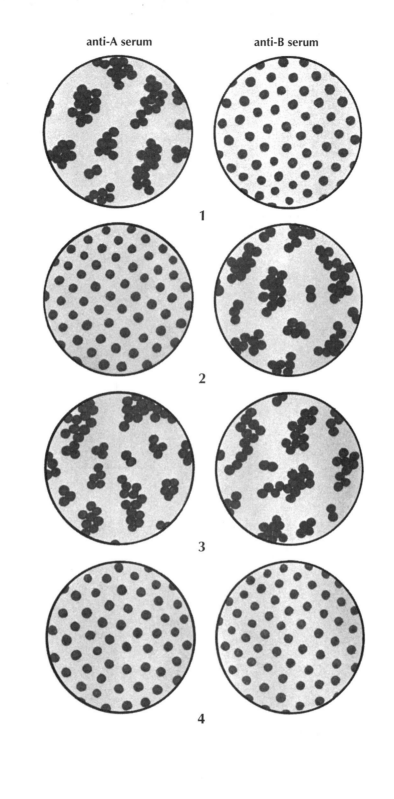

1

2

3

4

Blood typing

1. _____   3. _____

2. _____   4. _____

VI. True-False

For each question, write T for true and F for false in the blank to the left of each number. If a statement is false, correct it by replacing the underlined term and write the correct statement in the blanks below the question.

_____ 1. Blood from a person with type A blood will agglutinate with type B antiserum.

_____

_____

_____ 2. Type O blood contains antibodies to both A and B antigens

_____

_____

_____ 3. Lymphocytes and monocytes are agranular leukocytes.

_____

_____

_____ 4. The element that characterizes hemoglobin is iodine.

_____

_____

_____ 5. Band cells are immature neutrophils.

_____

_____

_____ 6. Substances that prevent blood clotting are called procoagulants.

_____

_____

VII. Completion Exercise

Write the word or phrase that correctly completes each sentence.

1. The gas that is necessary for life and that is transported to all parts of the body by the blood is called

_____

2. One waste product of body metabolism is carried to the lungs to be exhaled. This gas is known as

_____

3. Blood cells are formed in the _____

4. The number of different types of white blood cells is _____

5. Some monocytes enter the tissues, mature, and become active phagocytes. These cells are called _____

6. The process whereby red blood cells are clumped together in a reaction with a specific antiserum _____

7. The oxygen-carrying pigment in red blood cells is called _____

8. The watery fluid that remains after a blood clot is removed is called _____

9. The most important function of certain lymphocytes is to engulf disease-producing organisms by the process of _____

VIII. Practical Applications

Study each discussion. Then write the appropriate word or phrase in the space provided.

Group A

1. Ms. G sustained numerous deep gashes when she accidentally broke a glass shower door. One of the cuts bled copiously. In describing this type of bleeding, the doctor used the word _____

2. While the physician attended to the wound, the technician drew blood for typing and other studies. Ms. G's blood was found to agglutinate with both anti-A and anti-B serum. Her blood was classified as group _____

3. Among the available donors were some whose blood was found to be free of both A and B surface antigens. They were classified as having blood type _____

4. Further testing of Ms. G's blood revealed that it lacked the Rh factor. She was therefore said to be _____

5. If Ms. G were to be given a transfusion of Rh-positive blood, she might become sensitized to the Rh protein. In that event her blood would produce counteracting substances called _____

Group B

On the medical ward there were a number of patients who required extensive blood studies.

1. A 5-year-old boy had a history of frequent fevers and a tendency to bleed easily. His skin was pale and his heart rate rapid. The laboratory estimated the percentage of the different types of white cells in a blood smear, a test called a(n) _____

2. Mrs. C's history included rapid weight loss, constant thirst, and episodes of fainting. Tests showed the presence of excessive sugar, or glucose, in the blood. This symptom is described as _____

3. Mr. B, age 28, had a history of heart disease due to bacteria that caused dissolution (dissolving) of red blood cells. This disintegration of red blood cells is known as _____

4. Teen-age Joyce visited the prenatal clinic. She was noted to be listless and pale and anemia was suspected. A sample of her blood was separated in a centrifuge in order to measure the volume percentage of red cells. This test is a(n) _____

IX. Short Essays

1. Name some materials that are transported in the blood.

_____

_____

_____

_____

2. What kind of information can be obtained from blood chemistry tests?

_____

_____

_____

_____

Memmler, RL, Cohen, BJ, Wood, DL. *STUDY GUIDE FOR STRUCTURE AND FUNCTION OF THE HUMAN BODY*, 6/e, © 1996, Lippincott-Raven Publishers

The Heart

I. Overview

The ceaseless beat of the heart day and night throughout one's entire lifetime is such an obvious key to the presence of life that it is no surprise that this organ has been the subject of wonderment and poetry. When the heart stops pumping, life ceases. The cells must have oxygen, and it is the heart's pumping action that propels oxygenated blood to them.

In size, the heart has been compared to a closed fist. In location, it is thought of as being on the left side, although about one third of the heart is located to the right of the midline. The apex (point) of the triangular heart is definitely on the left. It rests on the *diaphragm,* the dome-shaped muscle that separates the thoracic cavity from the abdominopelvic cavity.

The heart has two sides, in which the aerated blood (higher in oxygen) and the unaerated blood (lower in oxygen) are kept entirely separate. So the heart is really a *double pump,* in which the two sides pump in unison. The right side pumps blood to the lungs to be oxygenated, and the left side pumps blood to all other parts of the body.

Each side of the heart is divided into two parts or *chambers,* which are in direct communication. The upper chamber or *atrium* on each side opens directly into the lower chamber or *ventricle. Valves* between the chambers keep the blood flowing forward as the heart pumps. The atria are the receiving chambers for blood returning to the heart. The two ventricles pump blood to all parts of the body. Because they pump more forcefully, their walls are thicker than the walls of the atria. The coronary arteries supply blood to the heart muscle or *myocardium.*

The heartbeat originates within the heart at the *sinoatrial (SA) node,* often called the pacemaker. Electrical impulses from the pacemaker spread over special fibers in the wall of the heart to produce contractions, first of the two atria and then of the two ventricles. After contraction, the heart relaxes and fills with blood. The relaxation phase is called *diastole* and the contraction phase is called *systole.* Together these two phases make up one *cardiac cycle.*

II. Topics for Review

A. Structure of the heart wall
    1. Endocardium
    2. Myocardium
    3. Epicardium
B. The pericardium
C. Anatomy of the heart
    1. Septum
    2. Chambers
    3. Valves
D. Blood circuits
    1. Pulmonary
    2. Systemic
E. The cardiac cycle
    1. Diastole
    2. Systole
F. The conduction system of the heart
    1. Sinoatrial node (pacemaker)
    2. Atrioventricular node
    3. Atrioventricular bundle and bundle branches
    4. Purkinje fibers
G. Heart rate
H. Normal and abnormal sounds
I. Instruments used in heart studies

III. Matching Exercises

Matching only within each group, write the answers in the spaces provided.

Group A

tricuspid value	interatrial septum	aortic valve
endocardium	myocardium	interventricular septum
mitral valve	epicardium	pulmonic semilunar valve

1. The membrane that forms the heart valves and lines the interior of the heart   _____

2. The heart muscle, the thickest layer in the heart wall   _____

3. The outermost layer of the heart   _____

4. The partition that separates the two upper chambers of the heart   _____

5. The partition between the two lower chambers of the heart   _____

6. The right atrioventricular valve   _____

7. The left atrioventricular valve, also called the bicuspid valve   _____

8. The valve that prevents blood on its way to the lungs from returning to the right ventricle   _____

9. The valve that prevents blood from returning to the left ventricle   _____

Group B

arteries   sinoatrial node   systole
ventricles   veins   diastole
atrioventricular node   atrioventricular bundle

1. The contraction phase of the cardiac cycle   _____

2. The resting period that follows the contraction phase of the cardiac cycle   _____

3. The pacemaker of the heart, located in the upper wall of the right atrium   _____

4. The lower chambers of the heart   _____

5. The mass of conduction tissue located in the septum at the bottom of the right atrium   _____

6. The group of conduction fibers between the AV node and the ventricles   _____

7. Vessels that carry blood from the heart to the tissues   _____

8. Vessels that carry blood from the tissues back to the heart   _____

Group C

stroke volume   tachycardia   atria
murmur   functional   cardiac output
myocardium   bradycardia

1. Term for a heart rate of greater than 100 beats per minute   _____

2. The upper chambers of the heart   _____

3. The tissue that is supplied by the coronary arteries   _____

4. The volume of blood pumped by each ventricle in 1 minute   _____

5. A heart rate of less than 60 beats per minute   _____

6. A sound that may result from a heart defect, such as the abnormal closing of a heart valve   _____

165

7. The amount of blood ejected from a ventricle with each beat _____

8. Adjective for a type of murmur that is not associated with abnormalities of the heart _____

Group D

| fluoroscope | echocardiography | electrocardiograph |
| extrasystole | stethoscope | catheter |

1. A thin tube that can be passed into a blood vessel _____

2. A method for studying the heart using sound waves _____

3. An instrument for recording the electric activity of the heart _____

4. An instrument that uses x-rays in examining deep structures _____

5. A premature heart beat _____

6. A simple instrument used by the physician for listening to sounds from within the patient's body _____

IV. Multiple Choice

Select the best answer and write the letter of your choice in the blank.

1. The special membranes between cardiac muscle cells are called   1. _____

   a. interfaces
   b. intercalated disks
   c. chordae tendineae
   d. foramen ovale
   e. coarctations

2. An average cardiac cycle lasts about   2. _____

   a. 8 seconds
   b. 5 seconds
   c. 1 minute
   d. 0.8 seconds
   e. 30 seconds

3. The parasympathetic nerve that supplies the heart is the   3. _____

   a. glossopharyngeal
   b. oculomotor
   c. vagus
   d. accessory
   e. trigeminal

4. The heart rate is increased by which division of the nervous system?   4. _____

   a. craniosacral
   b. sympathetic
   c. somatic
   d. parasympathetic
   e. sensory

5. Which of the following is <u>not</u> a part of the conduction system
   of the heart?   5. _____

   a. bundle of His
   b. right bundle branch
   c. Purkinje fibers
   d. calcium channel
   e. atrioventricular node

6. A large vessel that carries venous blood back into the right atrium is the   6. _____

   a. right coronary artery
   b. epicardium
   c. interatrial septum
   d. heart block
   e. coronary sinus

V. Labeling

For each of the following illustrations, write the name or names of each labeled part on the numbered lines on page 192.

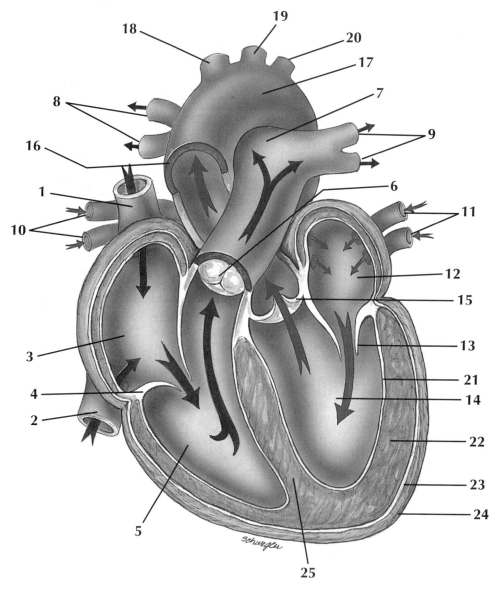

The heart and great vessels

1. _____

2. _____

3. _____

4. _____

5. _____

6. _____

7. _____

8. _____

9. _____

10. _____

11. _____

12. _____

13. _____

14. _____

15. _____

16. _____

17. _____

18. _____

19. _____

20. _____

21. _____

22. _____

23. _____

24. _____

25. _____

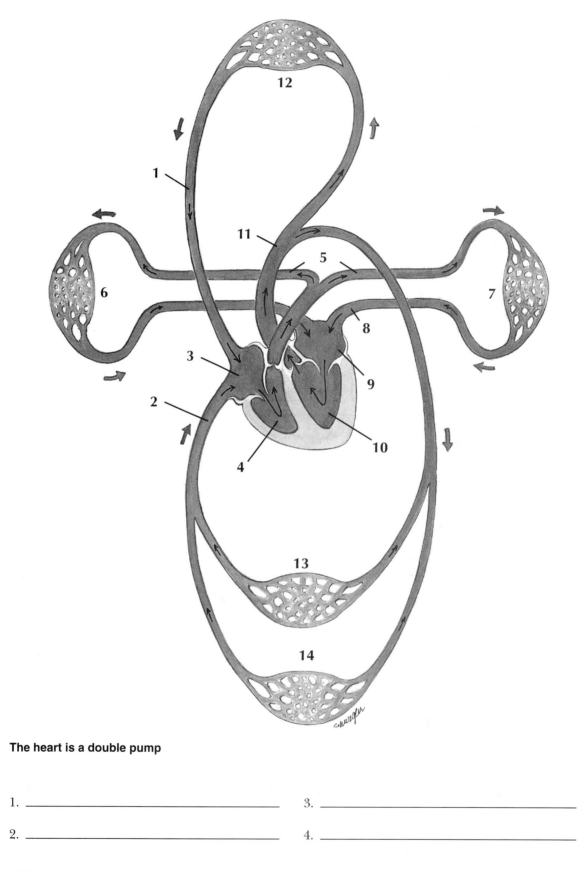

The heart is a double pump

1. _____   3. _____

2. _____   4. _____

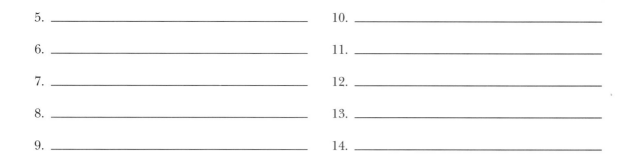

5. _____   10. _____

6. _____   11. _____

7. _____   12. _____

8. _____   13. _____

9. _____   14. _____

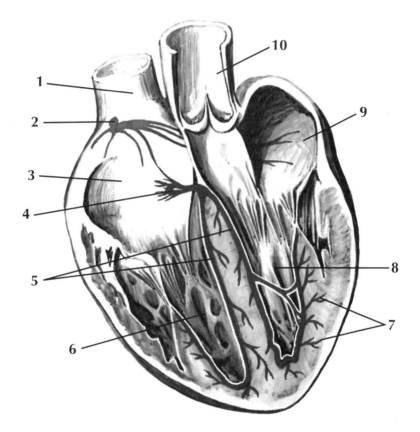

Conduction system of the heart

1. _____   6. _____

2. _____   7. _____

3. _____   8. _____

4. _____   9. _____

5. _____   10. _____

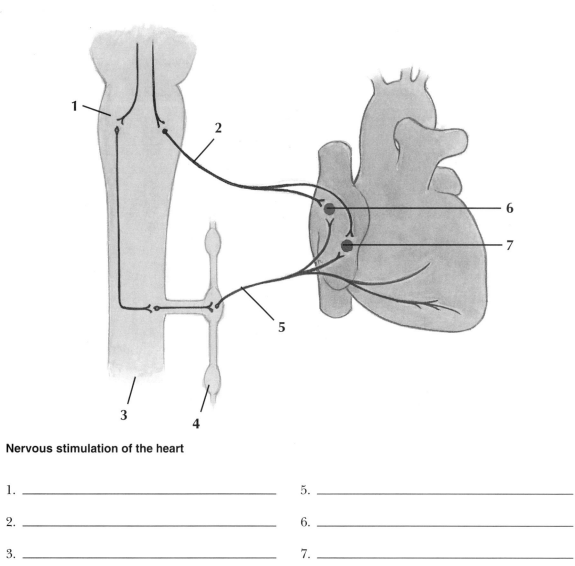

Nervous stimulation of the heart

1. _____   5. _____

2. _____   6. _____

3. _____   7. _____

4. _____

VI. True-False

For each question, write T for true and F for false in the blank to the left of each number. If a statement is false, correct it by replacing the <u>underlined</u> term and write the correct statement in the blanks below the question.

_____ 1. The right atrioventricular valve is the <u>bicuspid</u> valve.

_____

_____

_____ 2. The <u>systemic</u> circuit carries blood to the lungs.

_____

_____

_____ 3. The <u>epicardium</u> is the tissue that lines the interior of the heart.

_____

_____

_____ 4. A normal heart rhythm, or sinus rhythm, originates at the <u>sinoatrial</u> node.

_____

_____

_____ 5. When the ventricles contract, the atrioventricular valves are <u>closed</u>.

_____

_____

_____ 6. A rapid heart rate is described as <u>bradycardia</u>.

_____

_____

_____ 7. The vagus nerve is the <u>10th</u> cranial nerve.

_____

_____

VII. Completion Exercise

Write the word or phrase that correctly completes each sentence.

1. The fibrous sac that surrounds the heart is the   _____

2. The partition between the two sides of the heart is the   _____

3. Because each flap of the aortic and pulmonary valves is half-moon shaped, these valves are described as   _____

4. Another name for the left atrioventricular valve is the   _____

5. One complete sequence of relaxation and contraction of the heart is called a(n)

_____

6. The stroke volume and heart rate determine the volume of blood pumped by each ventricle in 1 minute. This volume is termed the

_____

7. The main influence over the heart rate outside of the heart itself is the

_____

8. A rapid, painless, and harmless method for studying the heart uses sound impulses that are reflected and recorded. This method is named

_____

VIII. Practical Applications

Study each discussion. Then write the appropriate word or phrase in the space provided.

Group A

1. Mrs. K had rheumatic fever several times during her teenage years. Now, in middle age, she is often short of breath and complains of spitting up blood. It is found that her left atrio-ventricular valve has become so scarred that blood cannot flow adequately from the left atrium to the left ventricle. The valve between the two left chambers of the heart is the

_____

2. Using a stethoscope, the physician listened to Mrs. K's heart. The first heart sound occurs during the contraction phase of the cardiac cycle. This active phase is called

_____

3. Mr. L was 42 years of age and overweight. During a game of handball he felt severe heart pains; he collapsed in shock. Examination indicated that a clot had formed in a blood vessel supplying the heart, with complete obstruction of blood flow. The vessels that supply the heart are the

_____

4. Mr. C, age 74, has not felt well for several months. He has had a long history of high blood pressure. Now he says that he feels weak, and he seems to be out of breath after slight exertion. Considering his history, his heart condition probably involves the thick, muscular layer of the heart wall, the

_____

5. One of the first tests that was done on all these patients was a recording of electrical currents produced by heart muscle. The apparatus that records this information is the

_____

Group B

1. Mr. A, age 61, has a history of heart disease. He now complains of dizziness, and his pulse rate is found to be 40. This abnormally low heart rate is termed

   _____

2. Further testing indicated that there was damage to the conduction system of Mr. A's heart. Normally the heartbeat begins at the pacemaker of the heart, which is technically called the

   _____

3. The greatest damage in the conduction system of Mr. A's heart was found in the bundle of His, which is also called the

   _____

IX. Short Essays

1. Although the heartbeat originates within the heart itself, it is influenced by factors in the internal environment. Describe some of these factors that can affect the heart.

   _____

   _____

   _____

   _____

2. Explain why the heart is described as a double pump.

   _____

   _____

   _____

   _____

Memmler, RL, Cohen, BJ, Wood, DL. *STUDY GUIDE FOR STRUCTURE AND FUNCTION OF THE HUMAN BODY*, 6/e, © 1996, Lippincott-Raven Publishers

CHAPTER

14

Blood Vessels and Blood Circulation

I. Overview

The blood vessels are classified, according to function, as **arteries, veins,** or **capillaries.** Arteries carry blood away from the heart; veins return blood to the heart. Small arteries are called **arterioles** and small veins are called **venules.** The walls of the arteries are thicker and more elastic than the walls of the veins, and the arteries contain blood under higher pressure. All vessels are lined with a single layer of simple epithelium called **endothelium.** The smallest vessels, the capillaries, are made only of this single layer of cells. It is through the walls of the capillaries that exchanges take place between the blood and the tissues. Materials move by diffusion as influenced by blood pressure and osmotic pressure.

The vessels carry blood through two circuits. The pulmonary vessels transport blood between the heart and the lungs for gas exchange. The systemic vessels distribute blood high in oxygen to all other body tissues and return deoxygenated blood to the heart.

The walls of the vessels, especially the small arteries, contain smooth muscle that is under the control of the involuntary nervous system. The diameters of the vessels can be regulated by the nervous system to alter blood pressure and to direct blood to various parts of the body as needed. These changes, termed **vasodilation** and **vasoconstriction,** are centrally controlled by a **vasomotor center** in the medulla of the brainstem.

Several forces work together to drive blood back to the heart in the venous system. The **pulse rate** and **blood pressure** are manifestations of the circulation; they tell the trained person a great deal about the overall condition of an individual.

177

II. Topics for Review

A. The blood vessels
   1. Structure and function
   2. Pulmonary and systemic circuits
B. Systemic arteries
   1. Branches of the aorta
   2. Branches of the iliac arteries
   3. Other parts of the arterial system
   4. Anastomoses
C. Systemic veins
   1. Superficial
   2. Deep
   3. Superior and inferior venae cavae
   4. Sinuses
D. The hepatic portal system
E. Circulation physiology
   1. Capillary exchanges
   2. Vasodilation and vasoconstriction
   3. Return of blood to the heart
F. Pulse
G. Blood pressure
   1. Contributing forces
   2. Measurement

III. Matching Exercises

Matching only within each group, write the answers in the spaces provided.

Group A

systemic circuit	endothelium	pulmonary circuit
artery	celiac trunk	carotid arteries
aorta	capillary	coronary arteries

1. The group of vessels that carries blood to and from the lungs for gas exchange   _____

2. A small vessel through which exchanges between the blood and the cells take place   _____

3. A blood vessel that carries blood away from the heart   _____

4. The tissue that comprises the innermost layer of a blood vessel   _____

5. The largest artery in the body   _____

6. The group of vessels that carries nutrients and oxygen to all tissues of the body except the lungs   _____

7. The vessels that branch off the ascending aorta and supply the heart muscle   _____

8. The vessels that supply the head and neck on each side   _____

9. A short unpaired artery that supplies some of the viscera of the upper abdomen

_____

Group B

blood pressure   portal system   anastomosis
valves   arteriole   vasomotor center
venule

1. A small artery

_____

2. A vessel that receives blood from the capillaries

_____

3. Term for a circuit that carries venous blood to a second capillary bed before it returns to the heart

_____

4. An area of the medulla that controls dilation and constriction of the blood vessels

_____

5. A communication between two arteries

_____

6. A force that drives materials out of the capillaries

_____

7. Structure that prevents blood from moving backward in the veins

_____

Group C

phrenic artery   lumbar arteries   common iliac arteries
brachiocephalic trunk   right subclavian artery   left common carotid artery
renal arteries   hepatic artery   superior mesenteric artery
brachial artery

1. A short artery that comes off the aortic arch and carries blood toward the head and the right arm

_____

2. The vessel that supplies the left side of the head and neck

_____

3. The vessel that carries oxygenated blood to the liver

_____

4. The largest branch of the abdominal aorta, a vessel that supplies most of the small intestine and the first half of the large intestine

_____

5. A vessel that supplies the diaphragm

_____

6. The branch of the brachiocephalic artery that supplies blood to the right upper extremity

_____

7. The main vessel that supplies the arm, a continuation of the axillary artery

_____

8. The large paired branches of the abdominal aorta that supply blood to the kidneys

_____

9. The vessels formed by final division of the abdominal aorta _____

10. The vessels that supply blood to the abdominal wall _____

Group D

brachiocephalic trunk radial artery   basilar artery
mesenteric arches   femoral artery volar arch
celiac trunk   circle of Willis

1. An anastomosis under the center of the brain formed by the
   two internal carotid arteries and the basilar artery _____

2. The anastomosis formed by the radial and ulnar arteries in
   the hand _____

3. Anastomoses between branches of the vessels supplying blood
   to the intestinal tract _____

4. The large vessel that branches into the right subclavian artery
   and the right common carotid artery _____

5. The short artery that branches into the left gastric artery,
   the splenic artery, and the hepatic artery _____

6. The vessel formed by union of the two vertebral arteries _____

7. The vessel in the thigh that is a continuation of the external
   iliac artery _____

8. The branch of the brachial artery that extends down the thumb
   side of the forearm and wrist _____

Group E

azygos vein   median cubital vein saphenous vein
inferior vena cava hepatic portal vein superior vena cava
jugular vein   brachiocephalic vein venous sinus

1. The longest vein _____

2. A vein frequently used for removing blood for testing because
   of its location near the surface at the front of the elbow _____

3. The vein that drains the area supplied by the carotid artery _____

4. The vessel formed by union of the jugular and subclavian veins _____

5. The vein that receives blood draining from the head, the neck,
   the upper extremities, and the chest _____

6. A vessel that drains blood from the chest wall and empties into
   the superior vena cava _____

7. A large channel that drains deoxygenated blood _____

8. The large vein that drains blood from the parts of the body below the diaphragm _____

9. The vein that receives blood from the unpaired abdominal organs and enters the liver _____

Group F

coronary sinus	hepatic veins	left testicular vein
common iliac veins	transverse sinuses	cavernous sinus
superior mesenteric vein	superior sagittal sinus	gastric veins

1. The two veins that unite to form the inferior vena cava _____

2. A paired vein that empties into the renal vein instead of emptying directly into the vena cava _____

3. The paired veins that drain the liver and empty directly into the inferior vena cava _____

4. The vein that drains most of the small intestine and the first part of the large intestine _____

5. The veins that drain the stomach and empty into the hepatic portal vein _____

6. The channel that receives blood from most of the veins of the heart wall _____

7. The channel that drains blood from the ophthalmic vein of the eye _____

8. The large lateral spaces between the layers of the dura mater that eventually receive nearly all the blood from the brain _____

9. A long blood-filled space in the midline above the brain and in the fissure between the two cerebral hemispheres _____

Group G

systolic	diastolic	sphygmomanometer
sinusoids	precapillary sphincter	pulse

1. A wave of increased pressure that begins at the heart when the ventricles contract and travels along the arteries _____

2. An instrument that is used to measure blood pressure _____

3. Term for the blood pressure reading taken during ventricular relaxation _____

4. Enlarged capillary channels where exchanges take place within the liver _____

5. Term for blood pressure measured during heart muscle contraction

_____

6. A circular structure that regulates blood flow into the tissues

_____

Group H

cerebral artery   femoral artery   radial artery
facial artery   brachial artery   saphenous vein
dorsalis pedis   coronary artery

1. A vessel that supplies the brain

_____

2. The artery that passes over the bone on the thumb side of the wrist and is often used to measure the pulse

_____

3. The artery on the top of the foot that is sometimes used for obtaining the pulse

_____

4. A vessel that supplies the heart muscle

_____

5. A large vessel from the thigh that often is used for a bypass graft

_____

6. A vessel that may be compressed against the lower jaw to stop hemorrhage around the nose and mouth

_____

7. The vessel that is compressed along the groove between the two large arm muscles to stop hemorrhage from the forearm, wrist, and hand

_____

8. The artery in the groin that is compressed to stop hemorrhage of the lower extremity

_____

IV. Multiple Choice

Select the best answer and write the letter of your choice in the blank.

1. Which of the following is <u>not</u> a subdivision of the aorta?   1. _____

     a. thoracic aorta
     b. descending aorta
     c. aortic arch
     d. pulmonary aorta
     e. abdominal aorta

2. Which of the following arteries is unpaired?   2. _____

     a. common carotid
     b. brachiocephalic
     c. external iliac
     d. brachial
     e. renal

3. Which of the following veins carries blood high in oxygen?   3. _____

    a. hepatic vein
    b. hepatic portal vein
    c. brachiocephalic vein
    d. superior vena cava
    e. pulmonary vein

4. The term *viscosity* means   4. _____

    a. solubility
    b. rate
    c. dilation
    d. thickness
    e. volume

5. The middle layer of the arterial wall is composed of elastic connective tissue and   5. _____

    a. cartilage
    b. endothelium
    c. smooth muscle
    d. adipose tissue
    e. skeletal muscle

6. As blood flows through the tissues, a force that draws fluid back into the capillaries is   6. _____

    a. blood pressure
    b. osmotic pressure
    c. hypertension
    d. vasoconstriction
    e. systolic pressure

V. Labeling

For each of the following illustrations, write the name or names of each labeled part on the numbered lines.

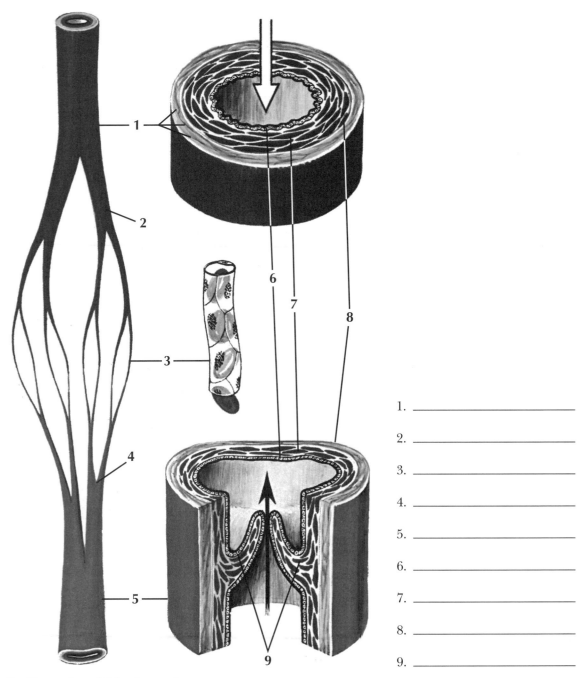

Sections of small blood vessels

1. _____
2. _____
3. _____
4. _____
5. _____
6. _____
7. _____
8. _____
9. _____

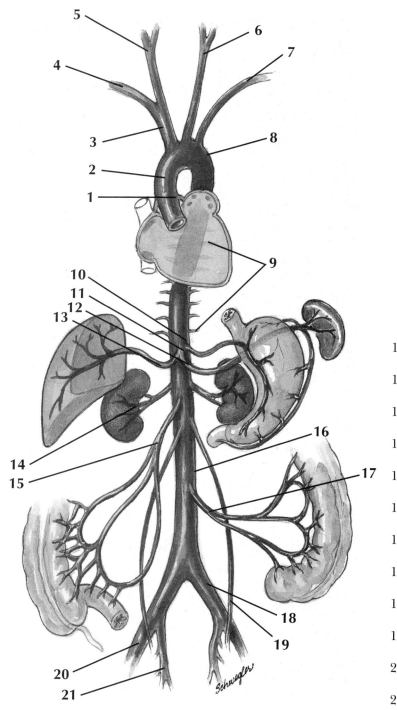
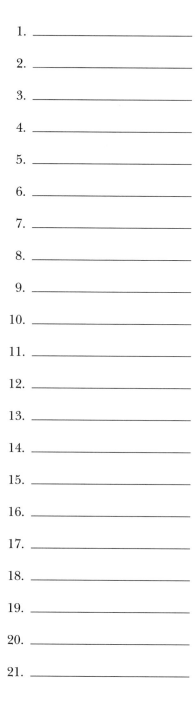

Aorta and its branches

1. _____
2. _____
3. _____
4. _____
5. _____
6. _____
7. _____
8. _____
9. _____
10. _____
11. _____
12. _____
13. _____
14. _____
15. _____
16. _____
17. _____
18. _____
19. _____
20. _____
21. _____

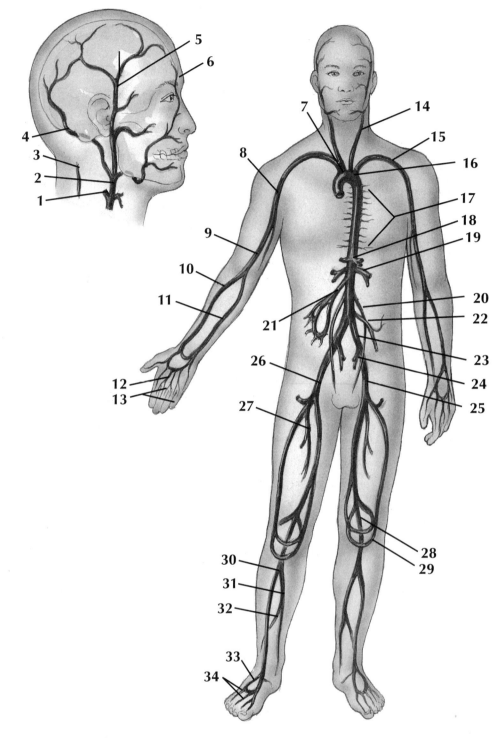

Principal systemic arteries

1. _____
2. _____
3. _____
4. _____
5. _____
6. _____
7. _____
8. _____
9. _____
10. _____
11. _____
12. _____
13. _____
14. _____
15. _____
16. _____
17. _____

18. _____
19. _____
20. _____
21. _____
22. _____
23. _____
24. _____
25. _____
26. _____
27. _____
28. _____
29. _____
30. _____
31. _____
32. _____
33. _____
34. _____

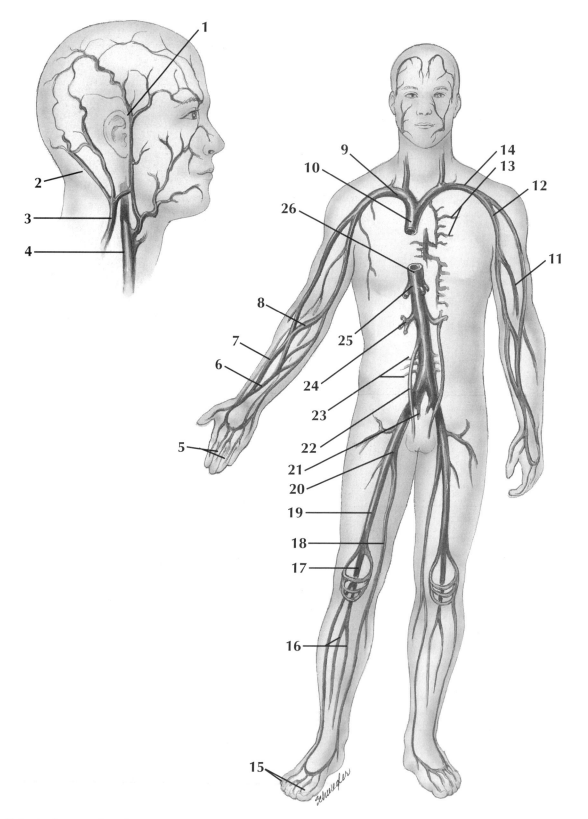

Principal systemic veins

1. _____

2. _____

3. _____

4. _____

5. _____

6. _____

7. _____

8. _____

9. _____

10. _____

11. _____

12. _____

13. _____

14. _____

15. _____

16. _____

17. _____

18. _____

19. _____

20. _____

21. _____

22. _____

23. _____

24. _____

25. _____

26. _____

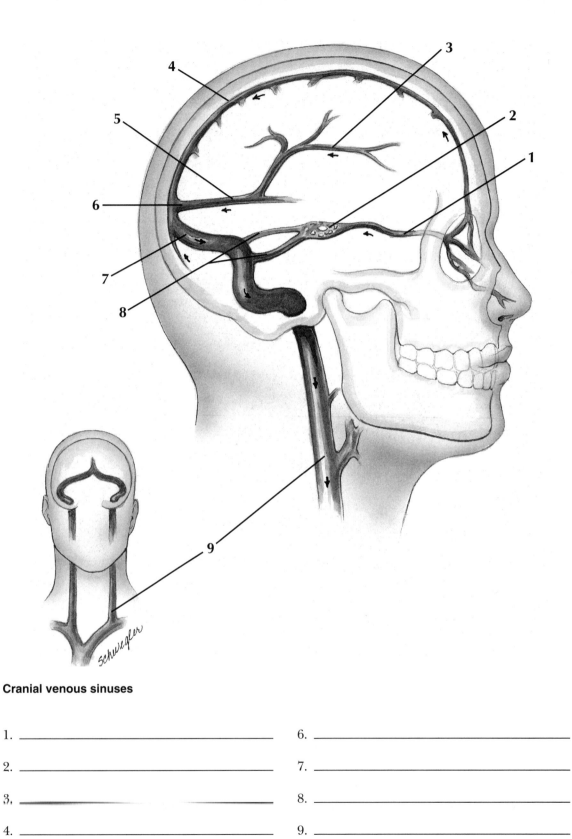

Cranial venous sinuses

1. _____

2. _____

3. _____

4. _____

5. _____

6. _____

7. _____

8. _____

9. _____

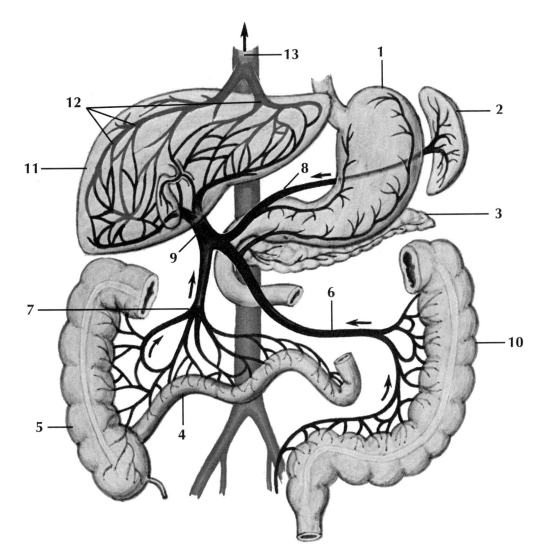

Hepatic portal circulation

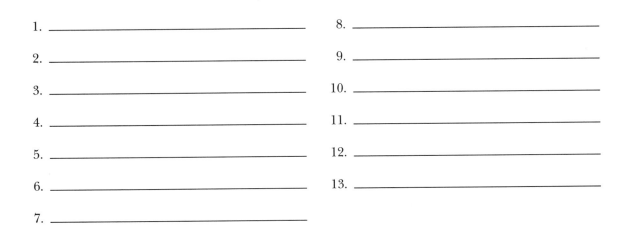

1. _____

2. _____

3. _____

4. _____

5. _____

6. _____

7. _____

8. _____

9. _____

10. _____

11. _____

12. _____

13. _____

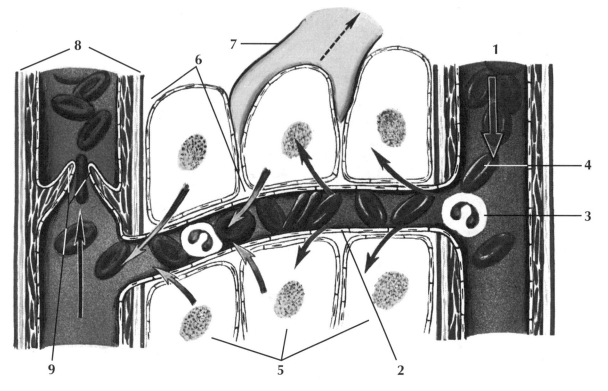

Diagram showing the connection between the small blood vessels through capillaries

1. _____   6. _____

2. _____   7. _____

3. _____   8. _____

4. _____   9. _____

5. _____

VI. True-False

For each question, write T for true and F for false in the blank to the left of each number. If a statement is false, correct it by replacing the <u>underlined</u> term and write the correct statement in the blanks below the question.

_____ 1. A small vein is called a <u>capillary</u>.

_____

_____

_____ 2. <u>Veins</u> carry blood toward the heart.

_____

_____

_____ 3. The widening of a blood vessel is termed <u>vasoconstriction</u>.

_____

_____

_____ 4. The circle of Willis is an anastomosis of vessels supplying blood to the <u>heart</u>.

_____

_____

_____ 5. The iliac veins drain blood from the <u>legs</u>.

_____

_____

_____ 6. <u>Diastolic</u> pressure is measured when the heart relaxes.

_____

_____

VII. Completion Exercise

Write the word or phrase that correctly completes each sentence.

1. Deoxygenated blood is carried from the right ventricle by the _____

2. The smallest subdivisions of arteries have thin walls in which there is little connective tissue and relatively more muscle. These vessels are _____

3. Supplying nutrients to all body tissues except the lungs and carrying off waste products from these tissues are functions of the vascular circuit described as _____

4. The innermost tunic of the vessels is composed of a special type of epithelium named _____

5. The longest veins in the body, located in the legs, are the _____

6. The smallest veins are formed by the union of capillaries. These tiny vessels are called _____

7. The circle of Willis is formed by a union of the internal carotid arteries and the basilar artery. Such a union of end arteries is called a(n) _____

8. One example of a portal system is the system that carries blood from the abdominal organs to the _____

VIII. Practical Applications

Study each discussion. Then write the appropriate word or phrase in the space provided.

Group A

1. Mr. S, age 53, complained of shortness of breath, weakness, and pain in the left chest. Examination indicated that the left semilunar valve was not functioning properly. This valve guards the entrance into the largest artery, which is the _____

2. Ms. M's job as a salesperson required her to stand for long hours. As a result she developed pain and swelling in the area of the long vein that extends up the medial side of each leg. This vein is the _____

3. Mr. K, age 71, was admitted to the hospital because he had fainted several times and was unable to recall events before and after these episodes. Physical examination showed a narrowing of the large artery on each side of the neck that carries blood to the brain. This artery is the _____

4. Small amounts of blood could still reach Mr. K's brain through the two vertebral arteries. These join at the base of the brain to form a single artery called the _____

5. Ms. K was protected from brain death by the fact that three arteries join to form an anastomosis under the brain. This anastomosis is the _____

Group B

1. Mr. J, an alcoholic, came in complaining of a variety of digestive and nervous disturbances. Examination revealed an enlarged liver and an accumulation of fluid in the abdominal cavity. There was evidence of obstruction of the large vein that drains the unpaired organs of the abdomen. This large vessel is called the _____

2. Ms. J and his digestive problems were studied for evidence of back pressure within the veins that drain the intestine. The larger of these is called the _____

3. Mrs. B, age 67, had been diabetic for several years. Now there was evidence of artery disease in several of the larger vessels. One area involved was the first portion of the aorta called the _____

4. Another vessel that was seriously damaged by hardening was the short one that is the first branch of the aortic arch. It is the _____

IX. Short Essays

1. Explain the purpose of vascular anastomoses.

_____

_____

_____

_____

2. What is the function of the hepatic portal system and what vessels contribute to this system?

_____

_____

_____

_____

3. Explain how the structure of the capillaries allows them to function in exchanges between the blood and the tissues.

_____

_____

_____

_____

Memmler, RL, Cohen, BJ, Wood, DL. *STUDY GUIDE FOR STRUCTURE AND FUNCTION OF THE HUMAN BODY*, 6/e, © 1996, Lippincott-Raven Publishers

The Lymphatic System and Immunity

I. Overview

Lymph is the watery fluid that flows within the lymphatic system. It originates from the blood plasma and from the tissue fluid that is found in the minute spaces around and between the body cells. The fluid moves from the *lymphatic capillaries* through the *lymphatic vessels* and then to the *right lymphatic duct* and the *thoracic duct*. These large terminal ducts drain into the subclavian veins, which carry blood back toward the heart. The lymphatic vessels are thin-walled and delicate; like some veins, they have valves that prevent backflow of lymph.

The *lymph nodes,* which are the system's filters, are composed of *lymphoid tissue.* These nodes remove impurities and manufacture *lymphocytes,* cells active in immunity. Chief among them are the cervical nodes in the neck, the axillary nodes in the armpit, the tracheo-bronchial nodes near the trachea and bronchial tubes, the mesenteric nodes between the peritoneal layers, and the inguinal nodes in the groin area.

In addition to these nodes, there are several organs of lymphoid tissue with somewhat different functions. The *tonsils* filter tissue fluid. The *thymus* is essential for development of the immune system during the early months of life. The *spleen's* numerous functions include destroying worn-out red blood cells, serving as a reservoir for blood, and producing red blood cells before birth.

Another part of the body's protective system is the *reticuloendothelial system,* which consists of cells involved in the destruction of bacteria, cancer cells, and other possibly harmful substances. Additional nonspecific defenses include the intact *skin* and *mucous membranes*, which serve as mechanical barriers, certain *reflexes*, such as sneezing and coughing, and the process of *inflammation*, by which the body tries to get rid of an irritant or to minimize its harmful effects.

The ultimate defense against disease is *immunity*, the means by which the body resists or overcomes the effects of a particular disease or other harmful agent.

There are two basic types of immunity: inborn and acquired. *Inborn immunity* is inherited; it may exist on the basis of *species, racial,* or *individual* characteristics. *Acquired immunity* is gained during an individual's lifetime. It involves reactions between foreign substances or *antigens* and the white blood cells known as *lymphocytes*. The *T cells* (T lymphocytes) respond to the antigen directly and produce *cell-mediated immunity*. *B cells* (B lymphocytes), when stimulated by an antigen, multiply into cells that produce specific *antibodies*, which react with the antigen. These circulating antibodies make up the form of immunity termed *humoral immunity*.

Acquired immunity may be *natural* (acquired before birth or by contact with the disease) or *artificial* (provided by a vaccine or an immune serum). Immunity that involves production of antibodies by the individual is termed *active immunity*; immunity acquired as a result of the transfer of antibodies to an individual from some outside source is described as *passive immunity*.

II. Topics for Review

A. The lymphatic system
 1. Lymphatic vessels
 2. Lymphoid tissue
  a. Lymph nodes
  b. Tonsils
  c. Thymus
  d. Spleen
B. Immunity
 1. Nonspecific defenses against invasion
  a. The reticuloendothelial system
  b. Skin and mucous membranes
  c. Phagocytosis
  d. Fever
  e. Inflammation
  f. Interferon

 2. Specific defenses: immunity
  a. Inborn
  b. Acquired
   (1) Active
   (2) Passive
 3. The immune response
  a. Lymphocytes
  b. Antigens
  c. Antibodies
 4. Vaccines and immunization

III. Matching Exercises

Matching only within each group, write the answers in the spaces provided.

Group A

right lymphatic duct   thymus   chyle
endothelium   inguinal nodes   axillary nodes
cervical nodes

1. The special single layer of cells that makes up the walls of
   lymphatic capillaries and blood capillaries   _____

2. A structure that is essential in the development of immunity
   very early in life   _____

3. The vessel that drains lymph from the right side of the head, the neck, the thorax, and the right upper extremity

_____

4. The milky-appearing fluid that is a combination of fat globules and lymph

_____

5. The lymph nodes located in the armpits

_____

6. The nodes that filter lymph from the lower extremities and the external genitalia

_____

7. The lymph nodes located in the neck that drain certain parts of the head and neck

_____

Group B

lacteals	spleen	palatine tonsils
lingual tonsils	pharyngeal tonsil	valves

1. The oval lymphoid bodies located at each side of the soft palate

_____

2. The mass of lymphoid tissue found on the back wall of the pharynx and commonly called adenoids

_____

3. Masses of lymphoid tissue at the back of the tongue

_____

4. Structures that prevent backflow of fluid in lymphatic vessels

_____

5. The organ that filters blood and is located in the upper left quadrant (left hypochondriac region) of the abdomen

_____

6. Specialized lymphatic capillaries of the intestine that absorb digested facts

_____

Group C

radial	thoracic duct	thymus
phagocytosis	antibodies	monocytes
subclavian vein		

1. The white blood cells that give rise to macrophages, cells active in the reticuloendothelial system

_____

2. Substances produced by lymphocytes that aid in combating infection

_____

3. The process by which cells engulf foreign substances, such as bacteria

_____

4. The organ in which T-lymphocytes mature

_____

5. The blood vessel on the right side of the body that receives lymph from the right lymphatic duct

_____

199

6. The large lymphatic vessel that drains lymph from below the diaphragm and from the left side above the diaphragm

_____

7. Term for the lymphatic vessels on the lateral side of the forearm

_____

Group D

| lymph | proteins | plasma |
| veins | superficial | cisterna chyli |

1. Term for lymphatic vessels located under the skin

_____

2. The temporary storage area formed by an enlargement of the first part of the thoracic duct

_____

3. Substances absorbed from tissue fluid into the lymph to be returned to the blood

_____

4. The liquid part of the blood that gives rise to intercellular fluid

_____

5. The fluid formed when tissue fluid passes from the intercellular spaces into the lymphatic vessels

_____

6. The vessels that often accompany the deep vessels of the lymphatic system

_____

Group E

| macrophages | hilum | chyle |
| thoracic duct | spleen | Kupffer's cells |

1. The area of exit for the vessels carrying lymph out of the node

_____

2. The name for monocytes that enter the body tissues and act to destroy foreign matter

_____

3. The phagocytes of the liver

_____

4. The fat-containing lymph that filters through the lacteals of the intestine

_____

5. The larger of the terminal vessels of the lymphatic system

_____

6. A lymphoid organ that produces red blood cells during embryonic and fetal life

_____

Group F

species immunity	racial immunity	B cells
active immunity	passive immunity	antigen
antibody	macrophages	

1. The type of protection that prevents humans from contracting certain animal diseases

_____

2. The type of protection illustrated by the greater resistance of black Americans over white Americans to malaria and yellow fever

_____

3. Any foreign substance introduced into the body that provokes an immune response

_____

4. The type of long-term immunity produced by infection or exposure to a microbial toxin

_____

5. The type of immunity that results from transfer of antibodies from mother to fetus through the placenta

_____

6. The lymphocytes that multiply in response to infection, giving rise to cells that produce antibodies

_____

7. Cells derived from monocytes that work with T cells

_____

Group G

| complement | attenuation | toxoid |
| plasma cell | immunization | gamma globulin |

1. Use of a vaccine to protect against infection

_____

2. The process of reducing the harmfulness of an organism to prepare a vaccine

_____

3. A toxin treated with heat or chemicals to reduce its harmfulness so that it may be used as a vaccine

_____

4. The fraction of the blood plasma that contains antibodies

_____

5. An antibody-producing cell derived from a B cell

_____

6. A group of blood proteins that may be needed to help an antibody destroy a foreign antigen

_____

IV. Multiple Choice

Select the best answer and write the letter of your choice in the blank.

1. Which of the following is <u>not</u> a type of cell associated with the reticuloendothelial system?

1. _____

   a. macrophages
   b. Kupffer's cells
   c. dust cells
   d. red blood cells
   e. monocytes

2. Thymosin is

2. _____

a. the hormone produced by the thymus gland
b. the fluid that drains from the intestine into the lymphatic system
c. the fluid in the lymphatic vessels
d. enlargement of the lymph nodes
e. the hormone produced by the thyroid gland

3. An organ that shrinks in size after puberty is the

3. _____

a. lymph node
b. lacteal
c. spleen
d. thymus
e. cisterna chyli

4. Which of the following is <u>not</u> a function of the spleen?

4. _____

a. phagocytosis
b. destruction of old red blood cells
c. splenectomy
d. filtration of blood
e. storage of blood

5. Which of the following is a specific defense against infection?

5. _____

a. skin
b. tears
c. mucus
d. antibodies
e. cilia

6. Cells that combine with foreign antigens and present them to the T cells are the

6. _____

a. allergens
b. B cells
c. macrophages
d. lymphocytes
e. viruses

7. Interleukins are

7. _____

a. a group of nonspecific proteins needed for agglutination
b. substances in the blood that react with antigens
c. the antibody fraction of the blood
d. substances released from macrophages that stimulate T cells
e. a type of immune serum

8. Which of the following terms does <u>not</u> describe a type of T cell?

8. _____

a. cytotoxic
b. suppressor
c. helper
d. memory
e. plasma

V. Labeling

For each of the following illustrations, write the name or names of each labeled part on the numbered lines.

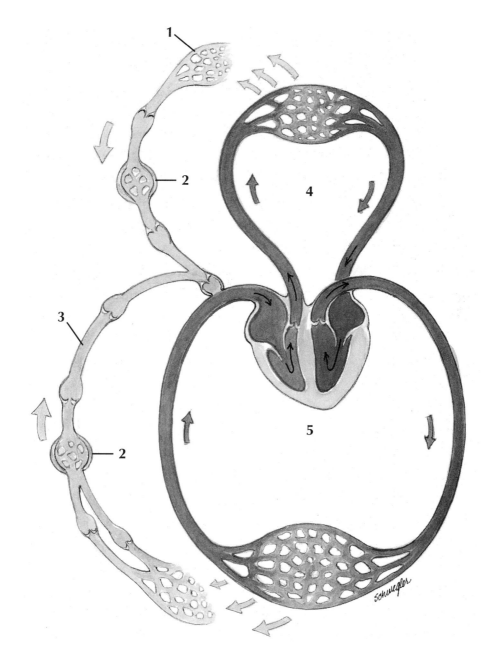

Lymphatic system in relation to the cardiovascular system

1. _____   4. _____

2. _____   5. _____

3. _____

1. _____

2. _____

3. _____

4. _____

5. _____

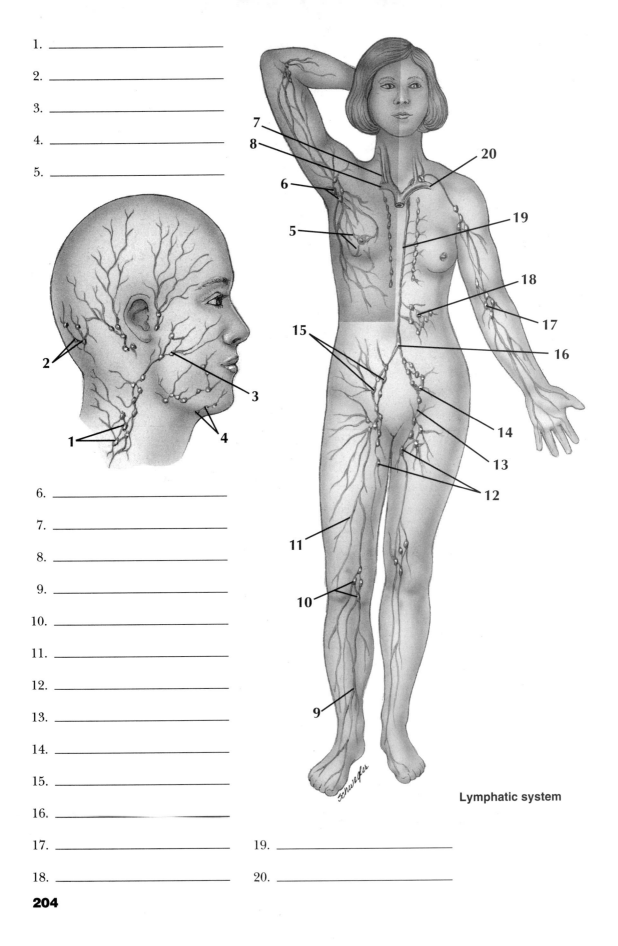

6. _____

7. _____

8. _____

9. _____

10. _____

11. _____

12. _____

13. _____

14. _____

15. _____

16. _____

17. _____

18. _____

19. _____

20. _____

Lymphatic system

VI. True-False

For each question, write T for true and F for false in the blank to the left of each number. If a statement is false, correct it by replacing the underlined term and write the correct statement in the blanks below the question.

_____ 1. The <u>superficial</u> lymphatic vessels are near the surface of the body.

_____

_____

_____ 2. The <u>inguinal</u> lymph nodes are in the groin.

_____

_____

_____ 3. The <u>thoracic</u> <u>duct</u> drains the lower part of the body and the upper left portion of the body.

_____

_____

_____ 4. The large lymphatic vessels empty into the <u>subclavian</u> <u>arteries</u>.

_____

_____

_____ 5. The <u>axillary</u> lymph nodes are located in the neck.

_____

_____

_____ 6. Adenoids is a common name for the <u>pharyngeal</u> <u>tonsil</u>.

_____

_____

_____ 7. The hormone produced by the thymus gland is <u>thymosin</u>.

_____

_____

_____ 8. The skin, mucous membranes, and secretions that destroy bacteria are examples of <u>nonspecific</u> defenses.

_____

_____

_____ 9. Immunity produced by transfer of antibodies from one person to another is described as <u>active</u> immunity.

_____

_____

_____ 10. The plasma cells that produce antibodies come from <u>T cells</u>.

_____

_____

_____ 11. Immunoglobulin is another name for <u>antibody</u>.

_____

_____

VII. Completion Exercise

Write the word or phrase that correctly completes each sentence.

1. The fluid that moves from tissue spaces into special collection vessels for return to the blood is called _____

2. Lymphatic vessels from the left side of the head, neck, and thorax empty into the largest of the lymphatic vessels, the _____

3. The lymph from the body below the diaphragm and from the left side above the diaphragm joins the bloodstream when the thoracic duct empties into the _____

4. Lymph nodes located between the two layers of peritoneum that form the mesentery are called _____

5. In city dwellers, nodes may appear black because they become filled with carbon particles. This is true mostly of the nodes that surround the windpipe and its divisions. These are the _____

6. The spleen and other organs produce cells that can engulf harmful bacterial and other foreign cells by a process called _____

7. Circulating antibodies are responsible for the type of immunity termed _____

8. There are two main categories of immunity. One is inborn immunity, whereas the other is _____

9. Antibodies transmitted from the mother's blood to the fetus provide a type of short-term borrowed immunity called _____

10. The administration of vaccine, on the other hand, stimulates the body to produce a longer-lasting type of immunity called _____

VIII. Practical Applications

Study each discussion. Then write the appropriate word or phrase in the space provided.

Group A

1. Mrs. B, age 52, was admitted for surgery. A biopsy of her breast mass was positive for cancer. Now she will undergo a mastectomy and removal of lymph nodes in the armpit. These nodes are called the

   _____

2. Mr. G, age 31, complained of swellings in his neck, his armpits, his groin, and other areas. A diagnosis of Hodgkin's disease was made. The nodes of the neck are designated the

   _____

3. Mr. J was 21 years old. His complaint concerned swelling in the groin region. A blood test showed that the young man had contracted syphilis. Infection of the external genitalia is often followed by enlargement of the lymph nodes in the groin, which are referred to as the

Group B

1. Mrs. R brought her 2-month-old infant to the office for the first of a series of injections to inoculate him against several serious diseases. The vaccine used for this purpose contains the weakened toxins of the organisms causing these diseases. A vaccine made from an altered toxin is known as a(n)

   _____

2. Ms. Y was allergic to pollens. In the hope that her tissues would become desensitized, the physician was giving her repeated injections of the substance that caused the reaction. The term for a foreign substance that produces an immune response is

   _____

3. Mr. N, age 42, had received a kidney transplant. He had appeared well for several months after the operation, when evidences of the rejection syndrome appeared. These reactions are due primarily to the activity of certain white blood cells that defend the body against foreign organisms. They are responsible for cell-mediated immunity and are called

   _____

4. Mr. S, a construction worker, was careless about cleansing his skin and neglected the abrasions on his fingers. An infection of his right thumb and index finger resulted in painful swelling and prevented him from working. He was treated with hot wet compresses, and in time his natural resistance overcame the infection. To a large extent, this resistance was due to the production of antibodies by the white cells known as

   _____

5. Mr. K consulted his physician for a generalized weakness and susceptibility to disease. Blood tests showed that he was infected with a virus that weakened his immune system by destroying helper T cells. The name of this disease is abbreviated by the letters _____

IX. Short Essays

1. Compare lymphatic capillaries and blood capillaries.

   _____

   _____

   _____

   _____

2. Describe the mechanisms that move lymph forward in the lymphatic vessels.

   _____

   _____

   _____

   _____

3. Compare nonspecific and specific defenses against disease and give examples of each type of defense.

   _____

   _____

   _____

   _____

UNIT

V

Energy: Supply and Use

16. RESPIRATION

17. DIGESTION

18. METABOLISM, NUTRITION, AND BODY TEMPERATURE

19. THE URINARY SYSTEM AND BODY FLUIDS

Memmler, RL, Cohen, BJ, Wood, DL. *STUDY GUIDE FOR STRUCTURE AND FUNCTION OF THE HUMAN BODY*, 6/e, © 1996, Lippincott-Raven Publishers

Respiration

I. Overview

Oxygen is taken into the body and carbon dioxide is released by means of the spaces and passageways that make up the ***respiratory system.*** This system contains the ***nasal cavities,*** the ***pharynx,*** the ***larynx,*** the ***trachea,*** the ***bronchi,*** and the ***lungs.***

Oxygen is obtained from the atmosphere and delivered to the cells by the process of ***respiration.*** The three phases of respiration are: ***pulmonary ventilation,*** normally accomplished by breathing; ***diffusion*** of gases between the alveoli of the lungs and the bloodstream; and ***transport*** of oxygen to the cells by the blood. Carbon dioxide is eliminated in a reverse pathway.

Oxygen is transported to the tissues almost entirely by the ***hemoglobin*** in red blood cells. Some carbon dioxide is transported in the red blood cells as well, but most is carried in the blood plasma as the ***bicarbonate ion.*** Carbon dioxide is important in regulating the pH of the blood and in regulating the breathing rate.

Breathing is primarily controlled by the ***respiratory control centers*** in the medulla and the pons of the brain stem. These centers are influenced by chemoreceptors located outside the medulla that respond to changes in the acidity of the cerebrospinal fluid. There are also ***chemoreceptors*** in the large vessels of the chest and neck that regulate respiration in response to changes in the composition of the blood.

II. Topics for Review

A. Phases of respiration
B. The respiratory system
C. Physiology of respiration
    1. Pulmonary ventilation
    2. Air movement
    3. Gas exchanges
    4. Gas transport
    5. Regulation of respiration
    6. Respiratory rates

III. Matching Exercises

Matching only within each group, write the answers in the spaces provided.

Group A

diffusion   inhalation   elasticity
exhalation   pulmonary ventilation   carbon dioxide
transport   respiration   surfactant

1. The total process by which oxygen is obtained from the environment and delivered to the cells _____

2. The gaseous waste product of cell metabolism _____

3. The exchange of air between the atmosphere and the air sacs of the lungs _____

4. The phase of respiration in which oxygen is carried to the cells by circulating blood _____

5. The first phase of respiration in which air is drawn into the lungs _____

6. The second phase of respiration in which air is expelled from the alveoli _____

7. The substance in the fluid lining the alveoli that prevents their collapse _____

8. The physical process by which a substance moves from an area where it is in higher concentration to an area where it is in lower concentration _____

9. The property that allows the lung and chest wall to recoil during exhalation _____

Group B

nasal septum   larynx   trachea
pharynx   conchae   nostrils
diaphragm   thyroid cartilage   sinus

1. The area below the nasal cavities that is common to both the digestive and respiratory systems _____

2. The scientific name for the cartilaginous structure commonly referred to as the voice box _____

3. The scientific name for the windpipe _____

4. The partition separating the two cavities of the nose _____

5. The three projections arising from the lateral walls of each nasal cavity

_____

6. The openings of the nose

_____

7. A small cavity in a bone of the skull lined with mucous membrane

_____

8. The muscle that separates the thoracic cavity from the abdominal cavity

_____

9. The structure that forms the "Adam's apple"

_____

Group C

hilum (or hilus)	epiglottis	chemoreceptors
bronchi	esophagus	vocal cords
cilia	oropharynx	nasopharynx

1. The hairlike structures that filter impurities within the conducting tubes of the respiratory tract

_____

2. The upper portion of the pharynx

_____

3. The portion of the pharynx located behind the mouth

_____

4. The tube leading from the pharynx that carries food into the stomach

_____

5. The structures that vibrate in the air flow from the lungs to aid in production of speech

_____

6. The leaf-shaped structure that helps to prevent the entrance of food into the trachea

_____

7. The two main air passageways to the lungs, formed by division of the trachea

_____

8. The notch or depression where the bronchus, blood vessels, and nerves enter the lung

_____

9. Areas in the major arteries of the chest and neck that regulate breathing according to changes in the composition of the blood

_____

Group D

diaphragm	pleura	bronchiole
glottis	mediastinum	carbonic anhydrase
alveoli	hemoglobin	carbon dioxide

1. The space between the two vocal cords

_____

2. The smallest division of a bronchus _____

3. The small air sacs in the lungs through which gases are
   exchanged _____

4. An enzyme in red blood cells that speeds the production of
   bicarbonate ion in the blood _____

5. The space between the lungs in which the heart and other
   structures are located _____

6. The structure that does most of the work of inhalation in quiet
   breathing _____

7. The serous membrane around each lung _____

8. The substance that carries most of the oxygen in the blood _____

9. The gas that yields bicarbonate ions when it dissolves in the
   blood _____

IV. Multiple Choice

Select the best answer and write the letter of your choice in the blank.

1. In respiration, gases move across epithelial membranes by the process of   1. _____

   a. filtration
   b. diffusion
   c. active transport
   d. carriers
   e. phagocytosis

2. Which of the following terms does <u>not</u> apply to the cells that line the
   conducting passages of the respiratory tract?   2. _____

   a. ciliated
   b. pseudostratified
   c. epithelial
   d. alveolar
   e. columnar

3. The partition in the nose is called the   3. _____

   a. septum
   b. pleura
   c. concha
   d. hilus
   e. glottis

4. The respiratory control centers are located in the parts of the brain stem called the

    a. cortex and spinal cord
    b. medulla and midbrain
    c. thalamus and pons
    d. pineal and ventricle
    e. pons and medulla

4. _____

5. The proportion of oxygen in inspired air is about

    a. 80%
    b. ⅔
    c. 21%
    d. ½
    e. 3.5%

5. _____

6. The amount of air moved into or out of the lungs in quiet, relaxed breathing is the

    a. residual volume
    b. total lung capacity
    c. vital capacity
    d. tidal volume
    e. functional residual capacity

6. _____

V. Labeling

For each of the following illustrations, write the name or names of each labeled part on the numbered lines.

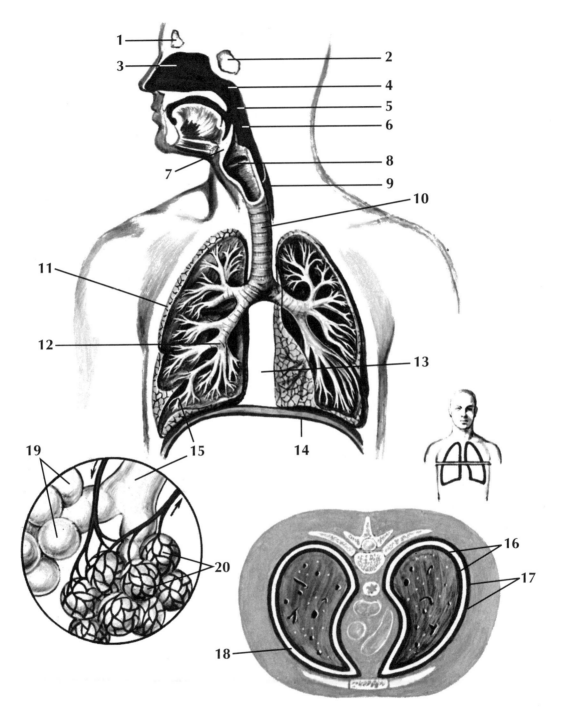

The respiratory tract

1. _____
2. _____
3. _____
4. _____
5. _____
6. _____
7. _____
8. _____
9. _____
10. _____

11. _____
12. _____
13. _____
14. _____
15. _____
16. _____
17. _____
18. _____
19. _____
20. _____

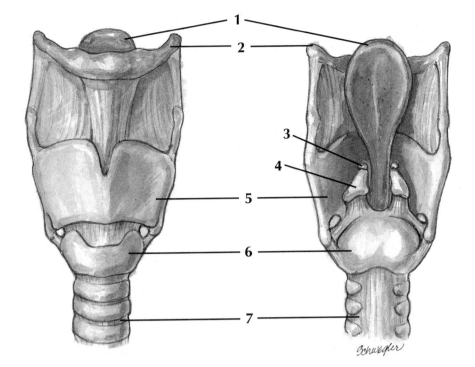

The larynx

1. _____   5. _____

2. _____   6. _____

3. _____   7. _____

4. _____

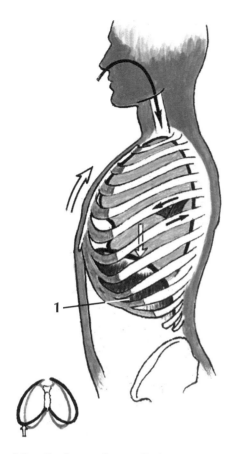
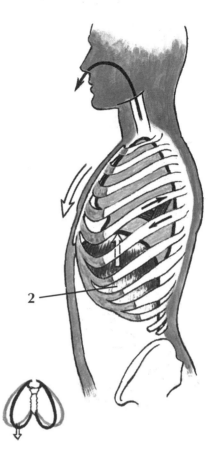

Action of the diaphragm in ventilation

1. _____   2. _____

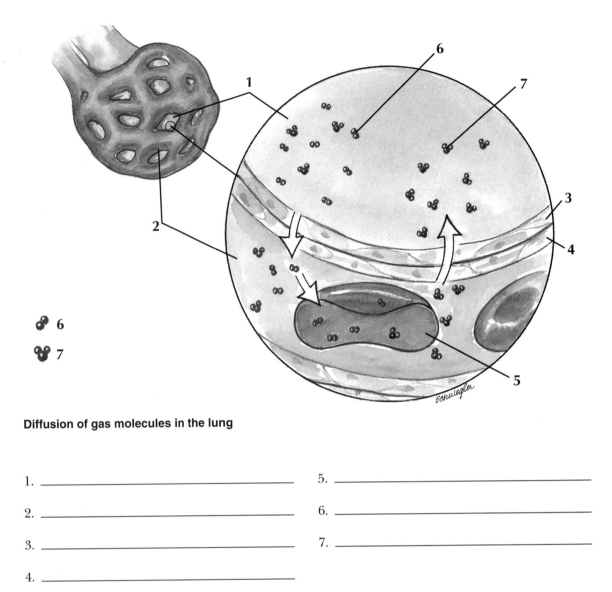

Diffusion of gas molecules in the lung

1. _____   5. _____

2. _____   6. _____

3. _____   7. _____

4. _____

VI. True-False

For each question, write T for true and F for false in the blank to the left of each number. If a statement is false, correct it by replacing the underlined term and write the correct statement in the blanks below the question.

_____ 1. The right bronchus divides into <u>three</u> secondary bronchi.

_____

_____

_____ 2. The portion of the pleura that is attached to the chest wall is the <u>visceral</u> pleura.

_____

_____

_____ 3. The wall of an alveolus is made of <u>stratified</u> squamous epithelium.

_____

_____

_____ 4. The <u>phrenic</u> nerve stimulates the diaphragm.

_____

_____

_____ 5. As carbon dioxide levels increase, the blood becomes more <u>alkaline</u>.

_____

_____

_____ 6. The amount of air that remains in the lung after maximum exhalation is the <u>residual</u> <u>volume</u>.

_____

_____

_____

VII. Completion Exercise

Write the word or phrase that correctly completes each sentence.

1. The main muscle of respiration is the   _____

2. The oxygen that diffuses into the capillary blood of the lungs is bound to a substance in red cells called   _____

3. The pons and medulla, which contain the respiratory control centers, are located in the part of the brain called the   _____

4. Much of the carbon dioxide in the blood is carried as an ion called   _____

5. The receptors in the carotid and aortic bodies that are involved in the control of respiration are of a type called   _____

6. The region that contains the heart, the large vessels, the esophagus, and the lymph nodes is the   _____

VIII. Practical Applications

Study each discussion. Then write the appropriate word or phrase in the space
provided.

Group A

1. Mr. C complained of a severe headache and facial pain. The
physician diagnosed the problem as an infection of the air
spaces within the cranial bones that are located near the
nasal cavities. These spaces are the   _____

2. G, age 14, complained of a sore throat and difficulty in swallowing. In describing the condition, the physician referred to the
throat by its scientific name of   _____

3. Ms. F, age 24, complained of hoarseness and said that it was
causing her difficulty in speaking to her students. She had an
inflammation of the structure that contains the vocal cords.
This is commonly called the voice box, but its scientific name is _____

4. Young D, age 5, had a profuse discharge from his nose. There
was inflammation of the nasal cavity lining. This membrane is
described specifically as a(n)   _____

5. Ms. F, age 24, complained of difficulty in breathing, partly
because of small growths called polyps which were forming between the small lateral projections at the side walls of the nasal
cavities. These three projections are the   _____

6. An additional problem for Ms. F was a deviated partition
between the two nasal spaces. This partition is the   _____

Group B

1. As a result of tuberculosis, Mrs. S, age 67, suffered from a painful condition that involved inflammation of the membrane
around the lung. This membrane is the   _____

2. Mrs. S's complaints included shortness of breath, a chronic
cough productive of thick mucus, and a "chest cold" of
2 months' duration. Her symptoms were due in part to the
obstruction and collapse of air sacs in the lungs. These millions
of tiny air sacs are the   _____

3. Evaluation of Mrs. S's respiratory function showed reduction in
the amount of air that could be moved into and out of her
lungs. The amount of air that can be expelled by maximum
exhalation following maximum inhalation is termed the   _____

4. Mr. G, age 47, was advised to see his doctor because a routine x-ray examination performed at his place of work revealed signs of lung cancer. This malignancy originated in one of the large passageways that carries air into the lungs. These tubes, subdivisions of the trachea, are the _____

IX. Short Essays

1. Name some parts of the respiratory tract where gas exchange does <u>not</u> occur.

   _____

   _____

   _____

   _____

2. Compare the different phases of breathing, indicating whether they are active or passive processes.

   _____

   _____

   _____

   _____

3. Explain briefly how respiration is regulated.

   _____

   _____

   _____

   _____

Memmler, RL, Cohen, BJ, Wood, DL. *STUDY GUIDE FOR STRUCTURE AND FUNCTION OF THE HUMAN BODY*, 6/e, © 1996, Lippincott-Raven Publishers

Digestion

CHAPTER 17

I. Overview

The food we eat is made available to cells throughout the body by the complex processes of *digestion* and *absorption.* These are the functions of the *digestive system;* its components are the *digestive tract* and the *accessory organs.*

The digestive tract, consisting of the *mouth,* the *pharynx,* the *esophagus,* the *stomach,* and the large and small *intestine,* forms a continuous passageway in which ingested food is prepared for use by the body and waste products are collected to be expelled from the body. The *liver,* the *gallbladder,* and the *pancreas,* the accessory organs, manufacture various enzymes and other substances needed in digestion. They also serve as storage areas for substances that are released as required.

Digestion begins in the mouth with the digestion of starch. It continues in the stomach, where proteins are digested, and is completed in the small intestine. Most absorption of digested food also occurs in the small intestine through small projections of the lining called *villi.*

The process of digestion is controlled by both nervous and hormonal mechanisms, which regulate the activity of the digestive organs and the rate at which food moves through the digestive tract.

II. Topics for Review

A. Wall of the digestive tract
B. The peritoneum
C. Organs of the digestive tract and their functions
D. Accessory organs and their functions
E. The process of digestion
    1. Enzymes
    2. Other substances
F. Absorption

G. Control of the digestive process
   1. Nervous
   2. Hormonal

III. Matching Exercises

Matching only within each group, write the answers in the spaces provided.

Group A

ingestion   absorption   mastication
digestion   deciduous   peritoneum
peristalsis

1. The process by which food is converted into substances small enough to be taken into the cells   _____

2. The transfer of digested food into the bloodstream   _____

3. The intake of food into the digestive tract   _____

4. The process of chewing   _____

5. The rhythmic motion that propels food along the digestive tract   _____

6. The largest serous membrane in the body   _____

7. Term that describes the baby teeth, based on the fact that they are lost   _____

Group B

incisors   parotid   submandibular
molars   canines   premolars

1. Another name for the eyeteeth   _____

2. The eight cutting teeth located in the front part of the oral cavity   _____

3. The largest of the salivary glands   _____

4. The grinding teeth located in the back part of the oral cavity   _____

5. The permanent teeth that replace the baby molars; the bicuspids   _____

6. Name of the salivary glands that are located near the body of the lower jaw   _____

Group C

uvula   duodenum   deglutition
amylase   esophagus   pharynx
sphincter   jejunum   rectum

1. The act of swallowing _____

2. The enzyme in saliva that digests starch _____

3. A circular muscle that acts as a valve _____

4. A fleshy mass that hangs from the soft palate _____

5. The scientific name for the throat _____

6. The tube that carries food into the stomach _____

7. The second part of the small intestine _____

8. The first part of the small intestine _____

9. The part of the large intestine between the sigmoid colon and the anus _____

Group D

anus   greater omentum   mesocolon
rugae   villi   mesentery
parietal   vermiform appendix   submucosa

1. Term that describes the portion of a serous membrane that lines a body cavity _____

2. An apron-like double membrane that extends downward from the stomach _____

3. The double-layered portion of the peritoneum that is attached to the small intestine _____

4. The distal opening of the digestive tract _____

5. The layer of connective tissue beneath the mucous membrane in the wall of the digestive tract _____

6. The section of the peritoneum that extends from the colon to the back wall of the abdomen _____

7. Folds that appear in the lining of the stomach when it is empty _____

8. The numerous tiny projections in the small intestine that greatly increase its absorbing surface _____

9. The small blind tube attached to the cecum _____

Group E

pyloric sphincter   ileocecal valve   lower esophageal sphincter
bolus   soft palate   epiglottis
chyme   defecation   ileum

1. The back portion of the oral cavity roof _____

2. A small portion of food mixed with saliva that is pushed into the pharynx in swallowing _____

3. A structure that covers the opening of the larynx during swallowing _____

4. The structure that guards the entrance into the stomach _____

5. The valve between the distal end of the stomach and the small intestine _____

6. The mixture that forms in the stomach when food is combined with gastric juice _____

7. The sphincter between the small and large intestine _____

8. The final, and longest, section of the small intestine _____

9. Elimination of the stool _____

Group F

glycogen	trypsin	albumin
urea	lipase	starch
carbohydrates	proteins	fats

1. The waste product manufactured in the liver that is later eliminated by the kidneys _____

2. A class of organic chemicals that includes sugars and starches _____

3. The type of food that is digested by bile _____

4. A plasma protein produced in the liver _____

5. The nutrient that is digested by pancreatic amylase _____

6. The type of food that is digested by gastric juice _____

7. The form in which glucose is stored in the liver _____

8. A pancreatic enzyme that splits proteins into amino acids _____

9. An enzyme that breaks fats into simpler compounds in digestion _____

Group G

liver	lacteals	feces
cecum	gallbladder	small intestine
hydrolysis	pepsin	colon

1. The organ from which most digested food is absorbed into the bloodstream

_____

2. An organ that stores nutrients and releases them as needed into the bloodstream

_____

3. The enzyme that digests proteins in the stomach

_____

4. The lymphatic capillaries in the villi of the small intestine that absorb digested fats

_____

5. The splitting of food molecules by the addition of water

_____

6. The small pouch at the beginning of the large intestine

_____

7. The accessory organ that stores bile

_____

8. The solid waste products of digestion

_____

9. The major portion of the large intestine

_____

IV. Multiple Choice

Select the best answer and write the letter of your choice in the blank.

1. Which of the following is <u>not</u> a portion of the peritoneum?

   1. _____

   a. mesocolon
   b. lesser omentum
   c. mesentery
   d. hiatus
   e. greater omentum

2. Which of the following is <u>not</u> associated with the accessory organs of digestion?

   2. _____

   a. liver
   b. vermiform appendix
   c. cystic duct
   d. pancreas
   e. common bile duct

3. The adjective *hepatic* refers to the

   3. _____

   a. stomach
   b. liver
   c. gallbladder
   d. ileum
   e. duodenum

4. The active ingredients in gastric juice are

    a. bile and trypsin
    b. amylase and pepsin
    c. pepsin and hydrochloric acid
    d. maltase and secretin
    e. sodium bicarbonate and GIP

4. _____

5. The sublingual salivary glands are located

    a. on the tongue
    b. under the tongue
    c. in the cheek
    d. in the oropharynx
    e. in front of the uvula

5. _____

6. Which of the following does <u>not</u> occur in the mouth?

    a. ingestion
    b. mastication
    c. moistening food
    d. digestion of starch
    e. absorption of food

6. _____

7. Which of the following is the correct order of tissue from the innermost to the outermost layer in the wall of the digestive tract?

    a. smooth muscle, serous membrane, mucous membrane, submucosa
    b. submucosa, serous membrane, smooth muscle, mucous membrane
    c. serous membrane, smooth muscle, submucosa, mucosa
    d. mucous membrane, submucosa, smooth muscle, serous membrane
    e. none of the above

7. _____

V. Labeling

For each of the following illustrations, write the name or names of each labeled part on the numbered lines.

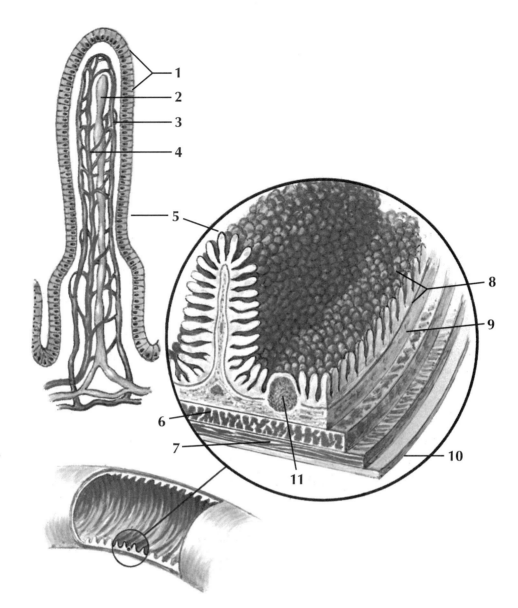

Wall of the small intestine

1. _____

2. _____

3. _____

4. _____

5. _____

6. _____

7. _____

8. _____

9. _____

10. _____

11. _____

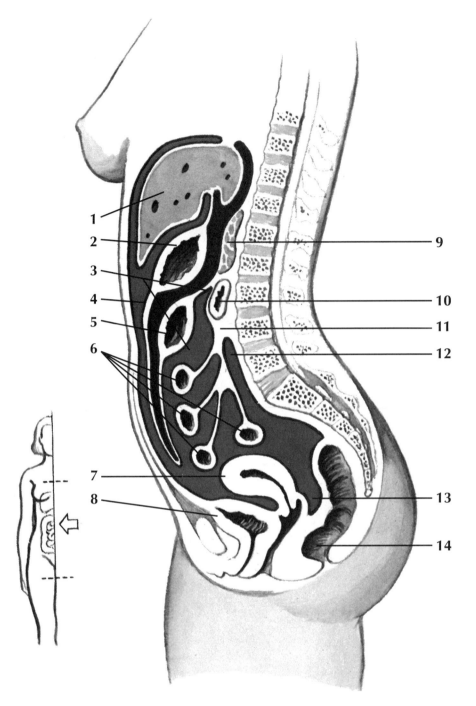

Abdominal cavity showing peritoneum

1. _____

2. _____

3. _____

4. _____

5. _____

6. _____

7. _____

8. _____

9. _____

10. _____

11. _____

12. _____

13. _____

14. _____

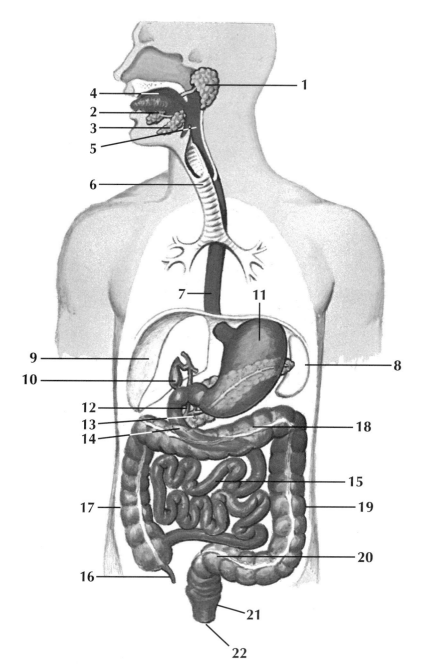

The digestive system

1. _____
2. _____
3. _____
4. _____
5. _____
6. _____
7. _____
8. _____
9. _____
10. _____
11. _____

12. _____
13. _____
14. _____
15. _____
16. _____
17. _____
18. _____
19. _____
20. _____
21. _____
22. _____

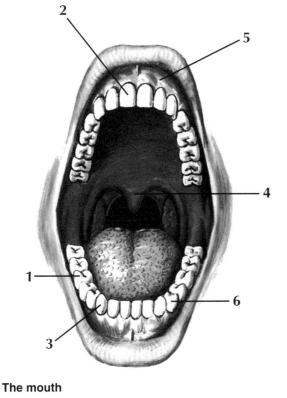

The mouth

1. _____
2. _____
3. _____
4. _____
5. _____
6. _____

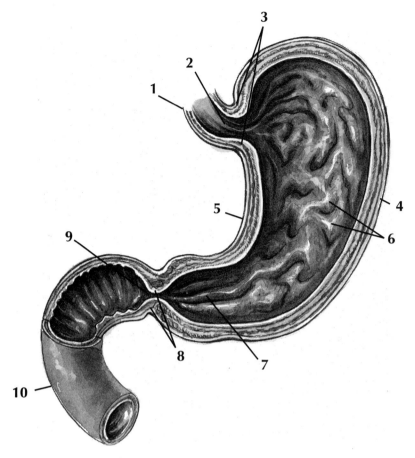

Longitudinal section of the stomach

1. _____   6. _____

2. _____   7. _____

3. _____   8. _____

4. _____   9. _____

5. _____   10. _____

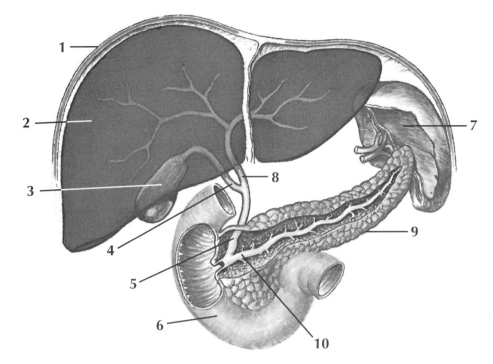

Accessory digestive organs and ducts

1. _____   6. _____

2. _____   7. _____

3. _____   8. _____

4. _____   9. _____

5. _____   10. _____

VI. True-False

For each question, write T for true and F for false in the blank to the left of each number. If a statement is false, correct it by replacing the <u>underlined</u> term and write the correct statement in the blanks below the question.

_____ 1. There are <u>20</u> deciduous teeth.

_____

_____

_____ 2. Most digested fats are absorbed into the <u>lymph</u>.

_____

_____

_____ 3. The mesocolon is the part of the peritoneum that is attached to the small intestine.

_____

_____

_____ 4. The lower esophageal sphincter is also called the pyloric sphincter.

_____

_____

_____ 5. The jejunum is the middle portion of the small intestine.

_____

_____

_____ 6. The transverse colon is the middle portion of the colon.

_____

_____

_____ 7. The common hepatic duct and the cystic duct merge to form the common bile duct.

_____

_____

_____ 8. The cystic duct drains bile from the gallbladder.

_____

_____

_____ 9. Gastrin is a hormone that stimulates the pancreas.

_____

_____

_____ 10. Amino acids are the building blocks of carbohydrates.

_____

_____

VII. Completion Exercise

Write the word or phrase that correctly completes each sentence.

1. Saliva is produced by three pairs of glands, of which the largest are the ones located near the angles of the jaw. These are the

   _____

2. One component of gastric juice kills bacteria and thus helps defend the body against disease. This substance is

   _____

3. Most of the digestive juices contain substances that cause the chemical breakdown of foods without entering into the reaction themselves. These catalytic agents are

   _____

4. Starches and sugars are classified as

   _____

5. The lower part of the colon bends into an S-shape, so this part is called the

   _____

6. A temporary storage section for indigestible and unabsorbable waste products of digestion is a tube called the

   _____

7. The anal canal, the distal portion of the large intestine, leads to the outside through an opening called the

   _____

8. The muscular sac in which bile is stored to be released as needed is called the

   _____

9. The serous membrane that lines the abdominal cavity and covers the abdominal organs is the

   _____

VIII. Practical Applications

Study each discussion. Then write the appropriate word or phrase in the space provided.

Group A

1. Mr. C, age 36, complained of pain in the "pit of the stomach." Ingestion of food seemed to provide some relief. The physician ordered x-ray studies, which indicated that the first part of the small intestine was involved. This short section is called the

   _____

2. Mr. C was a tense man who divided up his long working hours with coffee and cigarette breaks. After work, he would consume several cocktails before dinner. The x-ray studies showed abnormality in the lining of the stomach. This innermost layer of the organ is the

   _____

3. Three-month-old John was brought to the clinic by his mother because he had suffered several bouts of vomiting and could not retain food. The tentative diagnosis was a narrowing of the valve between the stomach and duodenum. This ringlike muscle is the _____

4. Mrs. D, age 41, complained of indigestion, belching, and abdominal pain. The pain in the upper right side of the abdomen and right shoulder became so severe that she reported to the emergency room. She was diagnosed as having a disorder involving the small sac under the liver that stores bile. This muscular pouch is the _____

Group B

1. Mr. K, age 24, complained of pain in his lower abdomen, in the right iliac region. Other symptoms and blood counts indicated that he was suffering from an inflammation of a wormlike appendage of the cecum. The name of this structure is the _____

2. Mr. K required immediate surgery to avoid rupture and the spread of infection to the serous membrane that links the abdominal cavity and covers most of the abdominal organs. This large membrane is the _____

3. Mr. S, age 51, complained of blood in his stool. Investigation revealed a tumor in the last part of the colon. This S-shaped section of the colon is called the _____

4. Mrs. M, age 34, also complained of seeing blood at the time of defecation. Examination revealed enlarged veins called hemorrhoids located in the distal part of the large intestine immediately following the rectum. This leads to the outside through an opening called the _____

5. Mr. G., age 28, had felt under par for some time. Studies showed that Mr. G. had a virus infection of the liver, a disease called hepatitis. Among the many functions of the liver that can be affected by liver disorders is the production of a digestive juice called _____

IX. Short Essays

1. Briefly explain how the process of digestion is regulated.

_____

_____

_____

_____

2. The hepatic portal system carries blood from the digestive organs to the liver. Explain why this is necessary.

_____

_____

_____

_____

3. Describe some features of the small intestine that increase the surface area for absorption of nutrients.

_____

_____

_____

_____

Memmler, RL, Cohen, BJ, Wood, DL. *STUDY GUIDE FOR STRUCTURE AND FUNCTION OF THE HUMAN BODY*, 6/e, © 1996, Lippincott-Raven Publishers

Metabolism, Nutrition, and Body Temperature

I. Overview

The nutrients that reach the cells following digestion and absorption are used to maintain life. All the physical and chemical reactions that occur within the cells make up *metabolism,* which has two phases: a breakdown phase, or *catabolism,* and a building phase, or *anabolism.* In catabolism, nutrients are oxidized to yield energy for the cells in the form of ATP. This process, termed *cellular respiration,* occurs in two steps: the first is anaerobic (does not require oxygen) and produces a small amount of energy; the second is aerobic (requires oxygen). This second step occurs within the mitochondria of the cells. It yields a large amount of the energy contained in the food plus carbon dioxide and water.

By the various pathways of metabolism, the breakdown products of food can be built into substances needed by the body. The *essential* amino acids and fatty acids cannot be manufactured internally and must be taken in with the diet. Minerals and vitamins are also needed in the diet for health. Since ingested food is the source of all nourishment for the body, a balanced diet should be followed and "food fads" should be avoided.

The rate at which energy is released from nutrients is termed the *metabolic rate.* It is affected by many factors including age, size, sex, activity, and hormones. Some of the energy in nutrients is released in the form of heat, which serves to maintain body temperature. A steady temperature of about 37°C (98.6°F) is maintained by several mechanisms.

Heat production is greatly increased during periods of increased muscular or glandular activity. Most heat loss occurs through the skin, with a smaller loss by way of the respiratory system and the urine and feces. The *hypothalamus* of the brain maintains the normal temperature in response to the temperature of the blood and information received from temperature receptors in the skin. Regulation occurs through vasodilation and vasoconstriction of the surface blood vessels, activity of the sweat glands, and muscle activity.

II. Topics for Review

A. Metabolism
    1. Catabolism and anabolism
    2. Metabolic rate
B. Nutrition
    1. Fats, proteins, and carbohydrates
    2. Essential amino acids and fatty acids
    3. Minerals and vitamins
    4. Balanced diet
C. Body temperature
    1. Heat production
    2. Heat loss
    3. Temperature regulation
        a. Role of the hypothalamus
    4. Normal body temperature

III. Matching Exercises

Matching only within each group, write the answers in the spaces provided.

Group A

glycogen	kilocalorie	catabolism
anaerobic	anabolism	mitochondria
amino acid	ATP	

1. The metabolic breakdown of nutrients for energy _____

2. Term that describes the first phase of cellular respiration because it does not require oxygen _____

3. A compound that stores energy in the cell _____

4. The cell organelles in which the aerobic steps of metabolism occur _____

5. The unit used to measure the energy in foods _____

6. The storage form of glucose _____

7. The metabolic building of simple compounds into substances needed by cells _____

8. A building block of proteins _____

Group B

glycerol   minerals   essential
oxidation   glucose   enzymes
saturated

1. The nutrient that is the main energy source for cells _____

2. Term for an amino acid that must be taken in as part of the diet _____

3. A component of fats _____

4. The chemical term for the breakdown of nutrients to release energy _____

5. The catalysts of metabolic reactions _____

6. Elements needed for proper nutrition _____

7. Term for fats that are solid at room temperature and are mostly from animal sources _____

Group C

calciferol   niacin   calcium
vitamin B_1 vitamin A   vitamin B_{12}
vitamin C   iron

1. Another name for thiamine, a deficiency of which will result in beriberi _____

2. The characteristic element in hemoglobin, the oxygen-carrying compound in the blood _____

3. The vitamin that prevents dry, scaly skin and night blindness _____

4. Another name for vitamin D, the vitamin required for normal bone formation _____

5. The vitamin that is also called ascorbic acid _____

6. The vitamin needed for blood cell formation that is found in meat, milk, and eggs _____

7. The vitamin found in yeast, meat, grains, and legumes, a deficiency of which will lead to pellagra _____

8. A mineral needed for proper bone development that is found in dairy products and vegetables _____

Group D

cellular respiration	constriction	hypothalamus
homeostasis	conduction	insulation
evaporation	circulation	

1. A series of metabolic reactions in which energy is released from nutrients; also described as oxidation _____

2. The tendency of body processes to maintain a constant state _____

3. A means for distributing heat throughout the body _____

4. The transfer of heat to the surrounding air _____

5. Prevention of heat loss _____

6. The chief heat-regulating center, located in the brain _____

7. The change that occurs in blood vessels of the skin if too much heat is being lost from the body _____

8. Heat loss resulting from the conversion of a liquid, such as perspiration, to a vapor _____

IV. Multiple Choice

Select the best answer and write the letter of your choice in the blank.

1. The amount of energy needed to maintain life functions while the body is at rest is   1. _____

   a. basal metabolism
   b. metabolic rate
   c. anabolic rate
   d. rate of conduction
   e. hyperthermia

2. Glycogen is stored in the   2. _____

   a. blood vessels and nerves
   b. heart and brain
   c. kidney and spleen
   d. bones and fat
   e. liver and muscles

3. Peas and beans are classified as   3. _____

   a. minerals
   b. grains
   c. cereals

d. legumes
e. lipids

4. An element that is required for normal nerve and muscle activity and is found in fruits, meats, seafood, and milk is

    4. _____

   a. glycerol
   b. potassium
   c. copper
   d. niacin
   e. pyridoxine

5. The largest amount of heat is produced in the body by

    5. _____

   a. muscles and glands
   b. cartilage and adipose tissue
   c. epithelium and blood
   d. nerves and tendons
   e. sense organs and lymphoid tissue

6. Which of the following is <u>not</u> an avenue for heat loss from the body?

    6. _____

   a. feces
   b. skin
   c. muscles
   d. expired air
   e. urine

V. True-False

For each question, write T for true and F for false in the blank to the left of each number. If a statement is false, correct it by replacing the <u>underlined</u> term and write the correct statement in the blanks below the question.

_____ 1. An aerobic reaction requires <u>oxygen</u>.

_____

_____

_____ 2. The building phase of metabolism is called <u>catabolism</u>.

_____

_____

_____ 3. Most heat loss in the body occurs through the <u>skin</u>.

_____

_____

_____ 4. The element nitrogen is found in all <u>proteins</u>.

_____

_____

_____ 5. It is recommended that more than half of the calories in the diet should come from <u>fats</u>.

_____

_____

_____ 6. Most oils are <u>unsaturated</u> fats.

_____

_____

_____ 7. When body temperature rises above normal, the blood vessels <u>constrict</u>.

_____

_____

_____ 8. A fan increases heat loss by the process of <u>convection</u>.

_____

_____

VI. Completion Exercise

Write the word or phrase that correctly completes each sentence.

1. All the chemical reactions that sustain life together make up   _____

2. The process of oxidizing nutrients within the cell for energy is termed   _____

3. Organic substances needed in small amounts in the diet are the   _____

4. A gland important in the control of the metabolic rate is the   _____

5. The most important heat-regulating center is a section of the brain called the   _____

6. Shivering is a way of increasing body heat by increasing the activity of the   _____

7. The diet should include foods that contain the nitrogenous compounds classified as

   _____

8. The normal range of body temperature in degrees Celsius is

   _____

9. The formula for converting Fahrenheit temperatures to Celsius is

   _____

10. Practice changing Fahrenheit to Celsius. Show the figures for changing 50°F and 70°F to Celsius.

   _____

11. Practice changing Celsius to Fahrenheit. Show the figures for changing 10°C and 25°C to Fahrenheit.

   _____

VII. Practical Applications

Study each discussion. Then write the appropriate word or phrase in the space provided.

Group A

1. Mrs. S, age 76, was hospitalized for a fracture of the femur caused by a fall. Her physician suspected that the break had resulted from a general weakening of the bones due to osteoporosis. This disorder, common in elderly women, is caused by a number of factors, including the dietary lack of a mineral found in dairy products. This mineral is

   _____

2. Mr. C, age 78, was accompanied by his daughter to visit his family physician. His daughter was concerned about his general state of health and marked weight loss within several months after the death of his wife. The doctor asked that Mr. C keep a record of his food intake for 2 weeks. Review of this record suggested that he was not eating properly and was suffering from a general lack of proper nutrients in his diet. The doctor described his condition as one of borderline

   _____

3. Young Mr. N, age 17, had placed himself on a strict vegetarian diet that included no animal products. He was not careful in planning his meals, however, and his family soon began to notice his loss of appetite, irritability, and susceptibility to disease. The school dietitian, when questioned by his mother, suggested that he was not getting the right balance of proteins, especially the building blocks of proteins that must be taken in with the diet, the

   _____

4. When Ms. R, age 15, went for her regular dental examination, the dentist noticed that her gums bled easily and that she had small cracks at the corners of her mouth. Brief questioning suggested that because of a lack of fruits and vegetables in her diet she was suffering from a lack of vitamins, especially vitamin C and a group of vitamins that includes thiamine and riboflavin, the

   _____

Group B

A physician working in a desert area of southeastern California saw a variety of cases during the course of a day.

1. A 6-year-old patient appeared apathetic and tired. His face was flushed and hot. On taking his temperature, the nurse found it to be 105°F. The physician took the child's history and examined him, then instructed his mother to give the child cool sponge baths and administer the prescribed medication. The cool water would draw heat from the body in changing from a liquid to a vapor, reducing body temperature by the process of   _____

2. Men working on a construction project complained of tiredness, nausea, a rapid pulse. They felt better after resting in the shade and drinking water and fruit juices. Their fluid losses had been caused mainly by increased activity of the   _____

3. Mr. K, age 69, was normally inactive. Today he had spent all afternoon repairing a fence. His wife found him wandering aimlessly outside, noted that his skin was red, hot, and dry and brought him to the doctor. His skin was red because the blood vessels there had been caused to   _____

4. On a cool fall day, Mr. J spent several hours walking with friends in the canyons of some nearby mountains. He was clad in shorts and a lightweight shirt. Now he was stumbling, his speech was hard to understand, and he complained of being sleepy. His heat-regulating center in the brain had been unable to sustain a normal body temperature. This center is the   _____

VIII. Short Essays

1. Summarize the dietary recommendations illustrated by the U.S. Department of Agriculture Food Guide Pyramid.

   _____

   _____

   _____

   _____

2. Explain briefly how body temperature is regulated.

   _____

   _____

   _____

   _____

The Urinary System and Body Fluids

I. Overview

The urinary system comprises two **kidneys,** two **ureters,** one **urinary bladder,** and one **urethra.** This system is thought of as the body's main excretory mechanism; it is, in fact, often called the **excretory system.** The kidney, however, performs other essential functions; it aids in maintaining water and electrolyte balance and in regulating the acid–base balance (pH) of body fluids. The kidneys also secrete a hormone that stimulates red blood cell production and an enzyme that acts to increase blood pressure.

The functional unit of the kidney is the **nephron.** It is the nephron that produces **urine** from substances filtered out of the blood through a cluster of capillaries, the **glomerulus.** Oxygenated blood is brought to the kidney by the **renal artery.** The arterial system branches through the kidney until the smallest subdivision, the **afferent arteriole,** carries blood into the glomerulus. Blood leaves the glomerulus by means of the **efferent arteriole** and eventually leaves the kidney by means of the **renal vein.** Before blood enters the venous network of the kidney, exchanges occur between the filtrate and the blood through the **peritubular capillaries** that surround each nephron.

From 50% to 70% of a person's body weight is water. This water serves as a solvent, a transport medium, and a participant in cellular metabolism. Dissolved in the water are a variety of substances, such as electrolytes, gases, nutrients, hormones, and waste products. Body fluids are distributed in two main compartments:

1. The *intracellular fluid compartment* within the cells
2. The *extracellular fluid compartment* located outside the cells.

There are also special compartments for the eye humors, cerebrospinal fluid, serous fluids, and synovial fluids.

The composition of intracellular and extracellular fluids is an important factor in homeostasis. These fluids must have the proper levels of electrolytes and must be kept at a constant pH. The kidneys are the main regulators of body fluids. Other factors that aid in regulation include hormones, buffers, and respiration. The normal pH of body fluids is a slightly alkaline 7.4. When regulating mechanisms fail to control shifts in pH, either *acidosis* or *alkalosis* results.

Water balance is maintained by constant intake and output of fluids, as controlled by the thirst center in the hypothalamus. Normally, the amount of fluid taken in with food and beverages equals the amount of fluid lost through the skin and the respiratory, digestive, and urinary tracts.

II. Topics for Review

A. The urinary system
1. Kidneys
2. Ureters
3. Bladder
4. Urethra
B. Renal function
1. Glomerular filtration
2. Tubular reabsorption
3. Tubular secretion
4. Concentration of the urine
   a. Role of ADH
C. Urine
D. Body fluids
1. Fluid compartments
   a. Intracellular fluid compartment
   b. Extracellular fluid compartment
   c. Special compartments
2. Water balance
   a. Intake
   b. Output
   c. Thirst center
3. Electrolytes
   a. Cations and anions
   b. Role of hormones
4. Acid–base balance (pH)
   a. Buffers
   b. Kidney function
   c. Respiration

III. Matching Exercises

Matching only within each group, write the answers in the spaces provided.

Group A

Bowman's capsule	nephrons	excretion
medulla	adipose capsule	fibrous capsule
retroperitoneal space	urine	

1. The process of removing waste products from the body _____

2. The microscopic functional units of the kidney _____

3. The fluid eliminated by the excretory system _____

4. The membranous connective tissue that encloses the kidney _____

5. A hollow bulb at the proximal end of the nephron _____

6. The area behind the peritoneum that contains the ureters and the kidneys _____

7. The crescent of fat that helps to support the kidney _____

8. The inner region of the kidney that contains the collecting tubules _____

Group B

collecting tubules   epithelium   filtration
renal pelvis   convoluted tubule   urea
cortex   glomerulus   tubular reabsorption

1. The cluster of capillaries within Bowman's capsule of the nephron _____

2. A coiled portion of a nephron _____

3. The process that returns useful substances in the filtrate to the bloodstream _____

4. The main tissue that makes up the kidney _____

5. The process by which substances leave the glomerulus and enter Bowman's capsule _____

6. The outer region of the kidney _____

7. The ducts that receive urine from the distal convoluted tubules of the nephrons _____

8. The funnel-shaped basin that forms the upper end of the ureter _____

9. The main nitrogenous waste product of the cells _____

Group C

electrolytes   micturition   peristalsis
glucose   urethra   hilus
calyces   protein   hydrogen ions

1. The rhythmic contractions that move urine along the ureter from the kidneys to the bladder _____

2. Another name for urination _____

3. The tube that carries urine from the bladder to the outside _____

4. The type of nutrient that always contains nitrogen _____

5. Term for the ions contained in urine _____

6. The simple sugar that appears in the urine in cases of diabetes mellitus _____

7. The area where the artery, the vein, and the ureter connect with the kidney _____

8. The cuplike extensions of the renal pelvis that collect urine _____

9. The substance that is adjusted by the kidney to regulate the pH of body fluids _____

Group D

erythropoietin	diffusion	specific gravity
trigone	renin	tubular secretion
aldosterone		

1. An indication of the amount of dissolved substances in the urine _____

2. The hormone released by the kidney that stimulates the bone marrow to produce red blood cells _____

3. The process that allows constant interchanges to occur between fluid compartments _____

4. The adrenal hormone that promotes the reabsorption of sodium _____

5. The triangle at the base of the bladder _____

6. The process by which the renal tubule actively moves substances from the blood into the nephron to be excreted _____

7. The enzyme produced by the kidney that acts to increase blood pressure _____

Group E

| intracellular | extracellular | juxtaglomerular apparatus |
| homeostasis | hypothalamus | buffer |

1. A substance that aids in maintaining a constant pH _____

2. Term that describes the fluid within the body cells _____

3. Term that describes fluids in spaces outside the body cells _____

4. The structure in the kidney that secretes renin _____

5. A constancy of internal conditions, such as the composition of body fluids _____

6. The part of the brain that controls the sense of thirst _____

Group F

| anion | cation | alkalosis |
| interstitial | parathyroid hormone | carbonic acid |

1. An ion with a positive electric charge _____

2. An ion with a negative electric charge _____

3. Term that describes fluid located in the microscopic spaces
between cells

_____

4. The substance formed when carbon dioxide goes into solution
in body fluids

_____

5. An increase in the pH of body fluids, such as may result from
hyperventilation

_____

6. A hormone that causes the kidney to reabsorb calcium

_____

IV. Multiple Choice

Select the best answer and write the letter of your choice in the blank.

1. Which of the following is <u>not</u> a function of the kidneys?

   a. adjust the composition of body fluids
   b. help regulate blood pressure
   c. regulate the volume of body fluids
   d. eliminate waste
   e. destroy red blood cells

1. _____

2. Select the correct order of urine flow from its source to the outside of
the body.

   a. bladder, renal pelvis, urethra, ureter
   b. kidney, ureter, bladder, urethra
   c. urethra, bladder, kidney, ureter
   d. bladder, ureter, cortex, nephron
   e. renal pelvis, nephron, urethra, calyx

2. _____

3. The juxtaglomerular apparatus consists of cells in the

   a. glomerulus and Bowman's capsule
   b. proximal convoluted tubule and efferent arteriole
   c. renal artery and cortex
   d. collecting tubules and renal vein
   e. distal convoluted tubule and afferent arteriole

3. _____

4. The enzyme renin raises blood pressure by activating

   a. red blood cells
   b. glomerular filtrate
   c. angiotensin
   d. erythropoietin
   e. sodium

4. _____

5. Which of the following is a normal constituent of urine?

   a. albumin
   b. glucose
   c. casts
   d. electrolytes
   e. blood

5. _____

6. The last part of each ureter enters at an angle through the lower bladder wall. This prevents

   a. reabsorption
   b. concentration of urine
   c. backflow of urine
   d. glomerular filtration
   e. micturition

6. _____

7. Which of the following fluids is <u>not</u> in the extracellular compartment?

   a. blood plasma
   b. interstitial fluid
   c. cytoplasm
   d. cerebrospinal fluid
   e. lymph

7. _____

8. Which of the following is <u>not</u> a route for water loss?

   a. food
   b. urine
   c. exhaled air
   d. skin
   e. feces

8. _____

V. Labeling

For each of the following illustrations, write the name or names of each labeled part on the numbered lines.

1. _____   10. _____

2. _____   11. _____

3. _____   12. _____

4. _____   13. _____

5. _____   14. _____

6. _____   15. _____

7. _____   16. _____

8. _____   17. _____

9. _____   18. _____

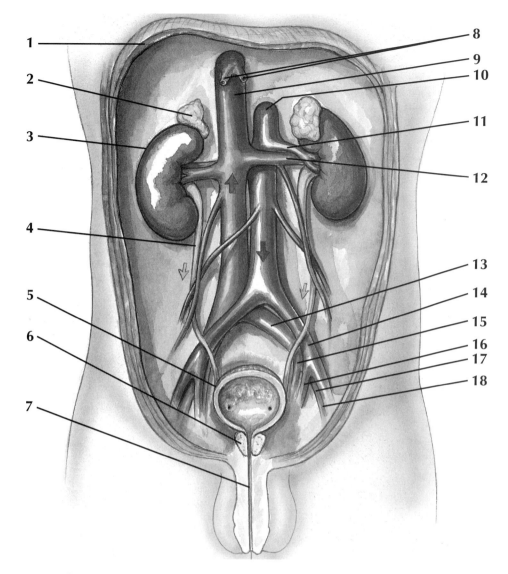

Urinary system with blood vessels

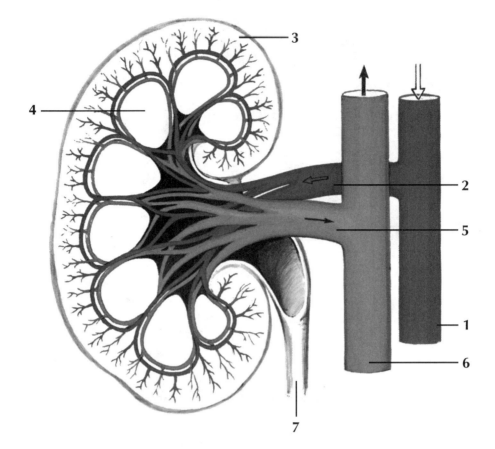

Blood supply and circulation of kidney

1. _____   5. _____

2. _____   6. _____

3. _____   7. _____

4. _____

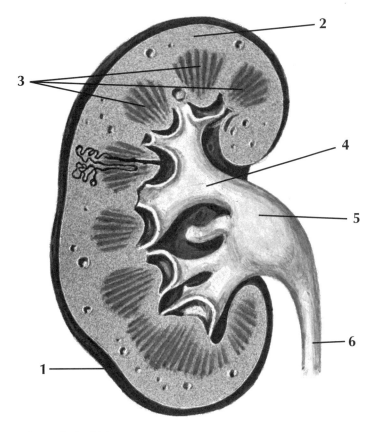

Longitudinal section through the kidney

1. _____   4. _____

2. _____   5. _____

3. _____   6. _____

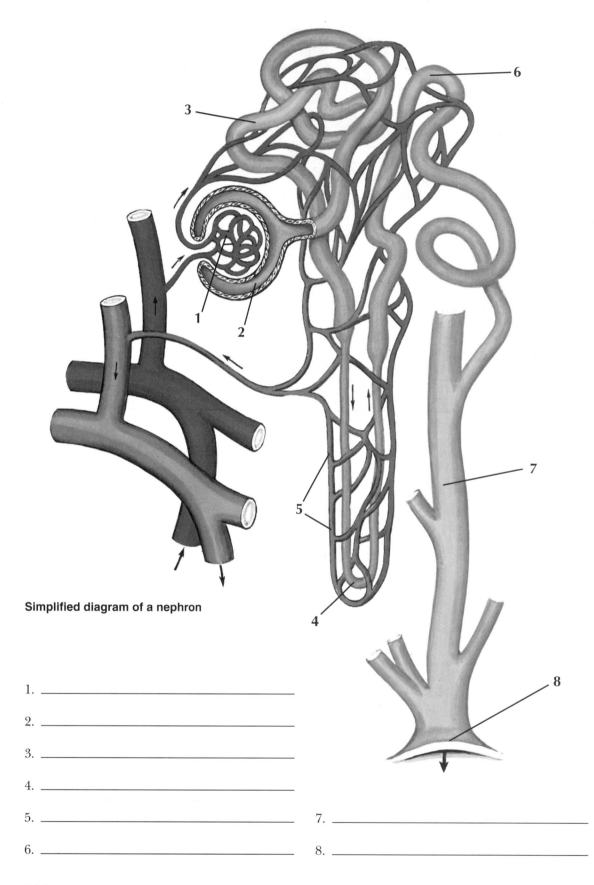

Simplified diagram of a nephron

1. _____

2. _____

3. _____

4. _____

5. _____

6. _____

7. _____

8. _____

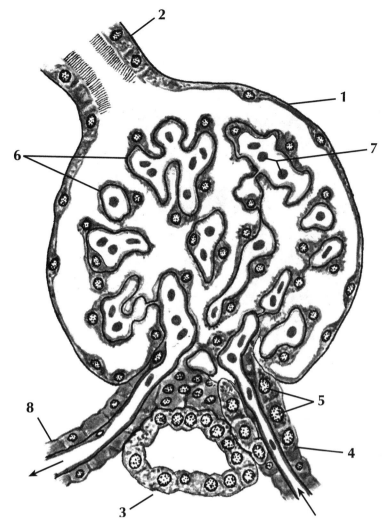

Structure of the juxtaglomerular apparatus

1. _____   5. _____

2. _____   6. _____

3. _____   7. _____

4. _____   8. _____

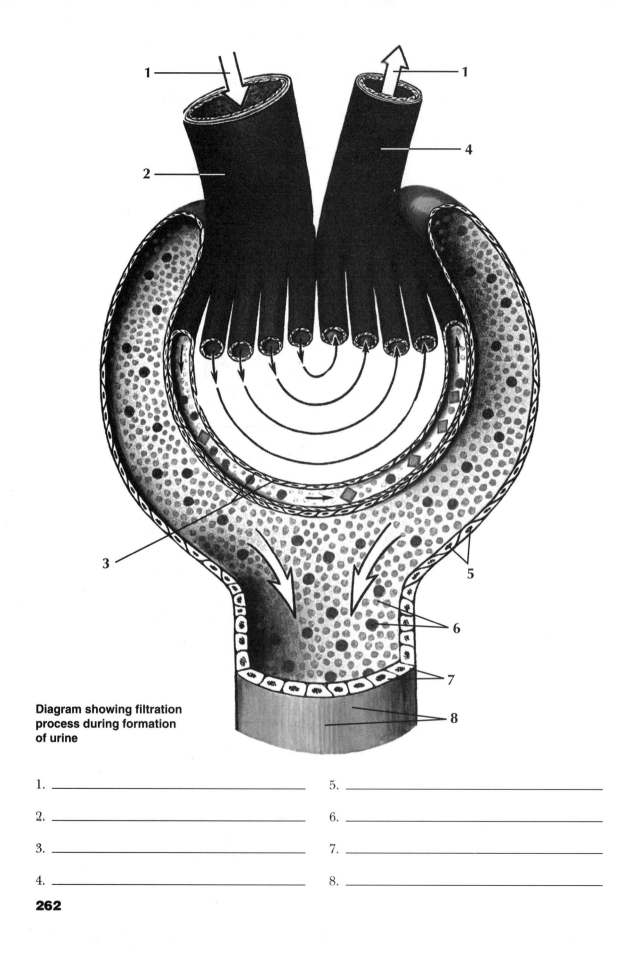

Diagram showing filtration process during formation of urine

1. _____   5. _____

2. _____   6. _____

3. _____   7. _____

4. _____   8. _____

262

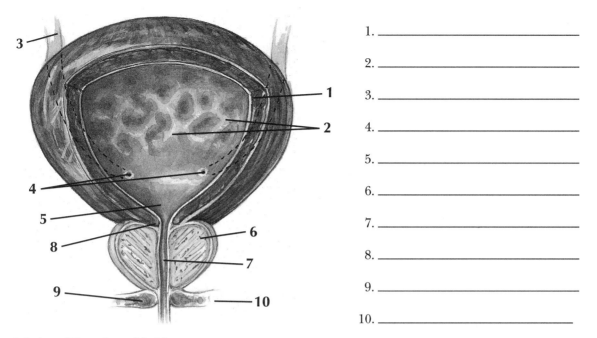

Interior of the urinary bladder

1. _____
2. _____
3. _____
4. _____
5. _____
6. _____
7. _____
8. _____
9. _____
10. _____

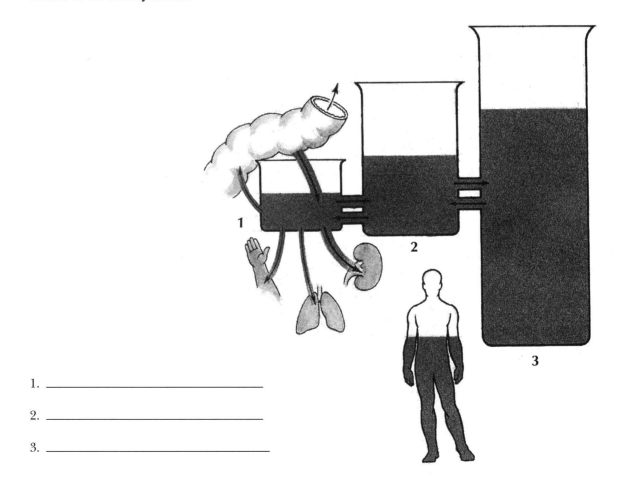

1. _____
2. _____
3. _____

The main fluid compartments

VI. True-False

For each question, write T for true and F for false in the blank to the left of each number. If a statement is false, correct it by replacing the underlined term and write the correct statement in the blanks below the question.

_____ 1. The inner portion of the kidney is the cortex.

_____

_____

_____ 2. Glomerular filtrate flows from Bowman's capsule into the proximal convoluted tubule.

_____

_____

_____ 3. Under the effects of ADH, water is reabsorbed.

_____

_____

_____ 4. The urethra transports urine from the kidney to the bladder.

_____

_____

_____ 5. Urine with a specific gravity of 0.04 is more concentrated than urine with a specific gravity of 0.02.

_____

_____

_____ 6. A substance with a pH above 7.0 is acidic.

_____

_____

_____ 7. A solution with a higher concentration than the fluid in the cell is termed hypotonic.

_____

_____

_____ 8. A solution of pH 4.0 is <u>more</u> acidic than a solution of pH 6.0

_____

_____

_____ 9. Sodium and potassium ions are positively charged, and thus are described as <u>cations</u>.

_____

_____

_____ 10. Body fluids are slightly <u>alkaline</u>.

_____

_____

_____ 11. The exhalation of carbon dioxide makes the blood more <u>acidic</u>.

_____

_____

VII. Completion Exercise

Write the word or phrase that correctly completes each sentence.

1. When the bladder is empty, its lining is thrown into the folds known as _____

2. The vessel that carries oxygenated blood to the kidney is the _____

3. The main constituent of urine is _____

4. The straddle type of injury occurs when, for example, a man is walking along a raised beam and slips so that the beam is between his legs. Such an accident may rupture the duct that transports urine away from the bladder. This duct is the _____

5. Diabetes insipidus is marked by great thirst and the elimination of large amounts of very dilute urine. The disease is caused by a lack of a hormone from the pituitary that regulates water reabsorption in the kidney. This hormone is _____

6. Water balance is partly regulated by a thirst center located in a region of the brain called the _____

7. Blood plasma, interstitial fluid, and lymph are contained in the fluid compartment outside the cells. This compartment is described as the

    _____

8. Body fluids are normally slightly alkaline at a pH of

    _____

9. The pH scale measures the concentration of

    _____

10. A drop in pH of body fluids produces a condition called

VIII. Practical Applications

Study each discussion. Then write the appropriate word or phrase in the space provided.

Group A

1. Mrs. L, age 31, was concerned because she had had several episodes of kidney disease during childhood. Tests demonstrated protein in her urine, indicating continuing damage to the working units of the kidneys, the

    _____

2. A test of Mrs. L's blood revealed an abnormally high content of the chief nitrogenous waste product called

    _____

3. Mrs. K was suffering from a bladder infection. Studies indicated that there was relaxation of the pelvic floor, causing stagnation of urine in the bladder, and corrective surgery was planned. In preparation for this, a catheter was inserted into the external opening, the

    _____

4. Mr. J, age 71, had difficulty emptying his bladder. The process of emptying the bladder is called urination or

    _____

5. One of the studies done in Mr. K's case revealed an obstruction at the bladder neck, a disorder that is fairly common in men of his age. The obstruction was caused by enlargement of the gland through which the first part of the urethra passes. This gland is the

    _____

6. Investigation of Mr. J's urinary problem included the passage of catheters up through the urethra and urinary bladder, and finally through the ureters into the kidney basin, also called the

    _____

Group B

1. Mr. M. was seen for a checkup, at which time a routine urinalysis showed the presence of glucose and ketones. The finding of these abnormal constituents in the urine is often an indication of the endocrine disorder known as

    _____

2. Mr. M's physician wanted to begin further evaluation right away. If this disorder goes untreated there can be a shift in the pH of body fluids as acid products accumulate. Substances in the blood that normally stabilize pH are call _____

3. Jane was suffering from Addison's disease, a deficiency of the adrenal cortex. Because this gland was not producing enough aldosterone, she was experiencing a loss of water and the mineral _____

IX. Short Essays

1. Name three systems other than the urinary system that excrete wastes. Specify the type of waste excreted by each.

   _____

   _____

   _____

   _____

2. Glomerular filtration is controlled by the permeability of the glomerulus. Name some substances that can pass through the glomerulus. Name some substances that cannot normally pass through the glomerulus.

   _____

   _____

   _____

   _____

3. Briefly describe the control center for the sense of thirst.

   _____

   _____

   _____

   _____

VI
UNIT

Perpetuating Life

20. THE MALE AND FEMALE REPRODUCTIVE SYSTEMS

21. DEVELOPMENT AND HEREDITY

Memmler, RL, Cohen, BJ, Wood, DL. *STUDY GUIDE FOR STRUCTURE AND FUNCTION OF THE HUMAN BODY*, 6/e, © 1996, Lippincott-Raven Publishers

The Male and Female Reproductive Systems

I. Overview

Reproduction is the process by which life continues. Human reproduction is *sexual,* that is, it requires the union of two different *germ cells* or *gametes.* (Some simple forms of life can reproduce without a partner in the process of *asexual* reproduction.) These germ cells, the *spermatozoon* in males and the *ovum* in females, are formed by *meiosis,* a type of cell division in which the chromosome number is reduced to one half. When fertilization occurs and the gametes combine, the original chromosome number is restored.

The reproductive glands or *gonads* manufacture the gametes and also produce hormones. These activities are continuous in the male but cyclic in the female. The male gonad is the *testis.* The remainder of the male reproductive tract consists of passageways for transport of spermatozoa; the male organ of copulation, the *penis;* and several glands that contribute to the production of *semen.* The female gonad is the *ovary.* The ovum released each month at the time of *ovulation* travels through the *oviducts* to the *uterus,* where the egg, if fertilized, develops. If no fertilization occurs, the ovum, along with the built-up lining of the uterus, is eliminated through the *vagina* as the *menstrual flow.*

Reproduction is under the control of hormones from the *anterior pituitary* which, in turn, is controlled by the *hypothalamus* of the brain. These organs respond to *feedback* mechanisms, which maintain proper hormone levels.

Aging causes changes in both the male and female reproductive systems. A gradual decrease in male hormone production begins as early as age 20 and continues throughout life. In the female, a more sudden decrease in activity occurs between ages 45 to 55 and ends in *menopause*, the cessation of menstruation and of the childbearing years.

II. Topics for Review

A. Formation of the germ cells
   1. Meiosis
   2. Gametes—ova and spermatozoa
B. The male reproductive tract
   1. Testes
   2. Ducts
   3. Penis
   4. Glands
      a. Semen
C. The female reproductive tract
   1. Ovaries
   2. Oviducts
   3. Uterus
   4. Vagina
   5. Vulva

D. Hormonal control of reproduction
   1. Pituitary
   2. Hypothalamus
   3. Feedback
E. The menstrual cycle
   1. Menopause
F. Contraception

III. Matching Exercises

Matching only within each group, write the answers in the spaces provided.

Group A

ovum   ovary   epididymis
testosterone   scrotum   seminiferous tubules
testis   spermatozoon

1. The specialized sex cell of the male   _____

2. The specialized sex cell of the female   _____

3. The female gonad   _____

4. The male gonad   _____

5. The sac suspended between the thighs that holds the testis   _____

6. The coiled ducts that comprise the bulk of the tissue in the testes   _____

7. The male hormone secreted by the interstitial cells of the testis   _____

8. The coiled tube in which spermatozoa are stored as they mature and become motile _____

Group B

ejaculatory duct   gamete   urethra
spermatic cord   gonad   semen
ductus deferens   FSH

1. A male or female sex cell _____

2. The tube that extends upward from the epididymis and transports spermatozoa _____

3. The structure containing the ductus deferens, nerves, blood vessels, and lymphatic vessels that extends from the testes on each side _____

4. The tube formed by the union of the ductus deferens and the duct from the seminal vesicle on each side _____

5. In males, the single tube that conveys urine and semen to the outside _____

6. The mixture of spermatozoa and glandular secretions that is expelled in ejaculation _____

7. The hormone that stimulates Sertoli cells in the testis _____

8. The general term for a sex gland _____

Group C

acrosome   ejaculation   prostate
penis   vasectomy   inguinal canal
seminal vesicle

1. The organ that holds the longest part of the urethra in the male _____

2. The channel through which the testes descend before birth _____

3. A gland located behind the urinary bladder in the male that contributes to the semen _____

4. A caplike covering over the head of the sperm cell that aids in penetration of the ovum _____

5. A procedure for sterilizing a male by removing a portion of the ductus deferens, thus preventing spermatozoa from reaching the urethra _____

6. A series of muscular contractions by which semen is expelled _____

7. The gland that contributes to semen and is located below the urinary bladder in the male _____

Group D

vulva
broad ligaments
ovulation

hymen
ovarian follicle
uterus

fimbriae
vagina

1. Peritoneal structures that serve as anchors for the uterus and ovaries _____

2. A sac within which the ovum matures _____

3. Discharge of an ovum from the surface of the ovary _____

4. Fringe-like extensions that sweep the ovum into the oviduct _____

5. The muscular organ in which a fetus develops _____

6. The muscular tube that serves as a birth canal _____

7. The external parts of the female reproductive system, consisting of the labia, the clitoris, and related structures _____

8. A fold of membrane found at the opening of the vagina _____

Group E

corpus
cervix
fallopian tubes

fundus
fornices

corpus luteum
endometrium

1. The small rounded part of the uterus located above the openings of the oviducts _____

2. The main part, or body, of the uterus _____

3. The neck-like part of the uterus that dips into the upper vagina _____

4. The specialized tissue that lines the uterus _____

5. Spaces formed as the uterus dips into the upper vagina, creating a circular recess _____

6. An alternate name for the oviducts _____

7. The structure formed by the ruptured follicle after ovulation _____

Group F

menstrual flow
estrogen
menopause

posterior fornix
clitoris

follicle stimulating hormone
luteinizing hormone

1. A hormone from the pituitary that causes the ovum to develop _____

2. The dorsal space behind the upper vaginal canal that lies next to the deepest part of the peritoneal cavity in the female _____

3. The ovarian hormone that begins the thickening of the endometrium early in the menstrual cycle _____

4. A small, highly sensitive organ in the vulva of the female _____

5. A hormone from the pituitary that causes rupture of the ovarian follicle and release of the ovum midway during the menstrual cycle _____

6. Cessation of reproductive activity in the female _____

7. The thickened lining of the endometrium that is shed when reproductive hormone levels decline _____

IV. Multiple Choice

Select the best answer and write the letter of your choice in the blank.

1. Which of the following is <u>not</u> part of the uterus?   1. _____

   a. fundus
   b. cervix
   c. fimbriae
   d. endometrium
   e. corpus

2. The tube that is cut in a vasectomy is the   2. _____

   a. epididymis
   b. ductus deferens
   c. testicular artery
   d. seminal vesicle
   e. urethra

3. The gland through which the first part of the urethra passes in males   3. _____

   a. prostate
   b. Cowper's gland
   c. Bartholin's gland
   d. bulbourethral gland
   e. seminal vesicle

4. A tubal ligation is done to   4. _____

   a. remove the uterus
   b. promote fertilization
   c. sterilize a male
   d. sterilize a female
   e. increase sperm count

5. The pelvic floor in both males and females is called the

5. _____

   a. cervix
   b. fornix
   c. epididymis
   d. perineum
   e. acrosome

V. Labeling

For each of the following illustrations, write the name or names of each labeled part on the numbered lines.

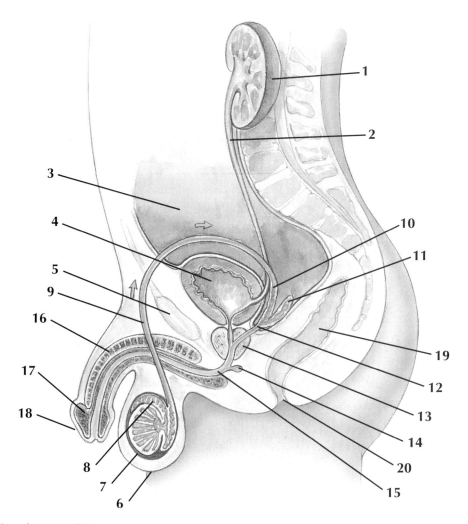

Male genitourinary system

1. _____

2. _____

3. _____

4. _____

5. _____

6. _____

7. _____

8. _____

9. _____

10. _____

11. _____

12. _____

13. _____

14. _____

15. _____

16. _____

17. _____

18. _____

19. _____

20. _____

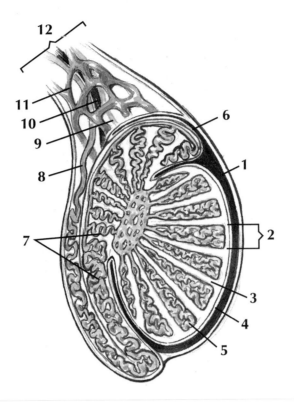

The testes

1. _____   7. _____

2. _____   8. _____

3. _____   9. _____

4. _____   10. _____

5. _____   11. _____

6. _____   12. _____

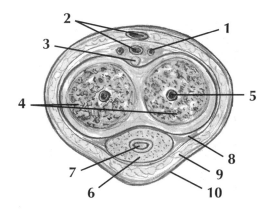

Cross section of the penis

1. _____

2. _____

3. _____

4. _____

5. _____

6. _____

7. _____

8. _____

9. _____

10. _____

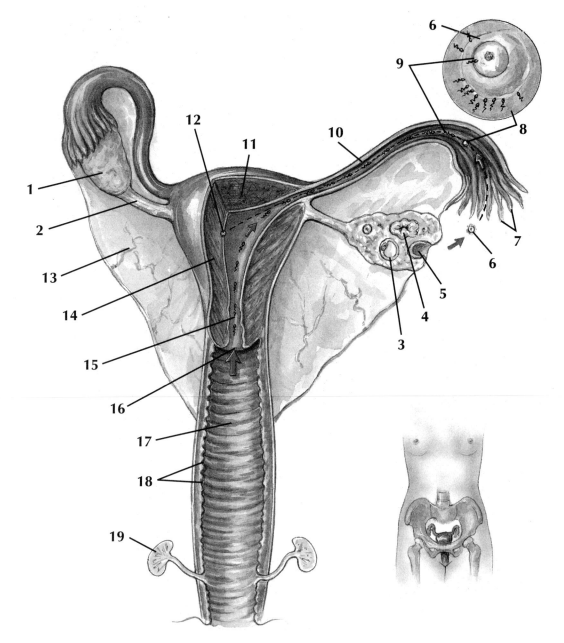

Female reproductive system

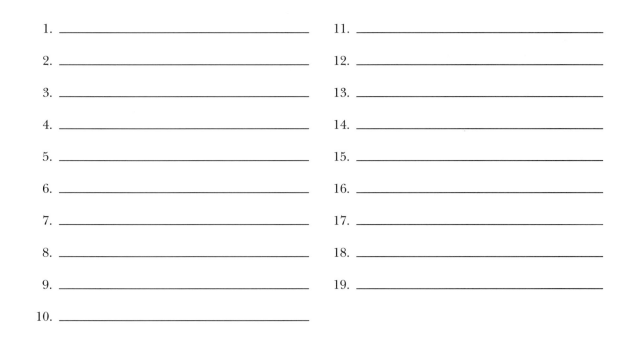

1. _____

2. _____

3. _____

4. _____

5. _____

6. _____

7. _____

8. _____

9. _____

10. _____

11. _____

12. _____

13. _____

14. _____

15. _____

16. _____

17. _____

18. _____

19. _____

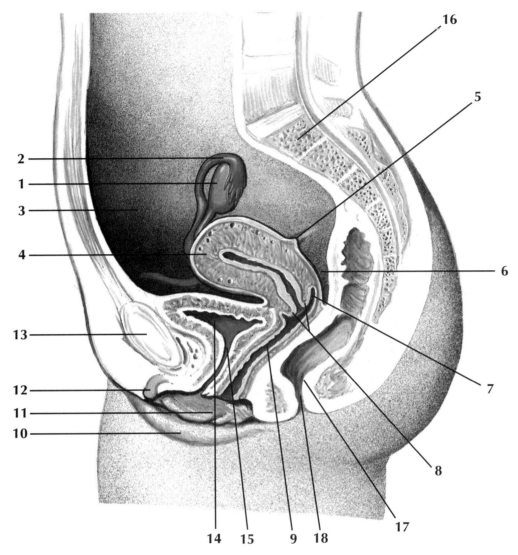

Female reproductive system, sagittal section

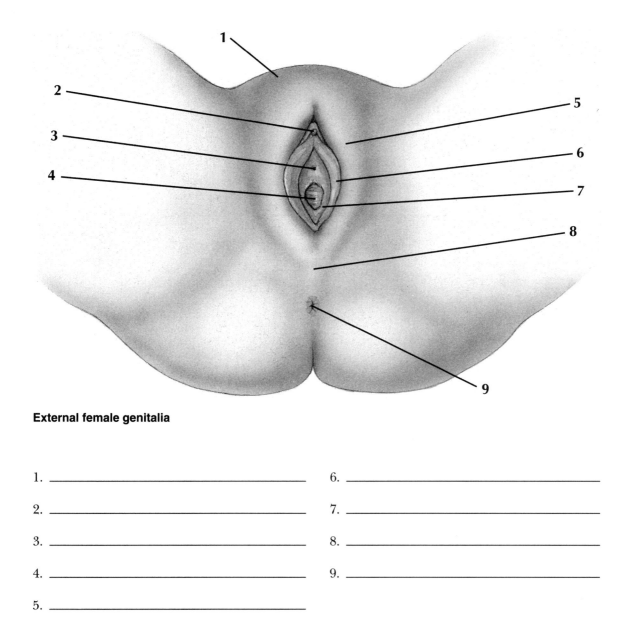

External female genitalia

1. _____   6. _____

2. _____   7. _____

3. _____   8. _____

4. _____   9. _____

5. _____

VI. True-False

For each question, write T for true and F for false in the blank to the left of each number. If a statement is false, correct it by replacing the <u>underlined</u> term and write the correct statement in the blanks below the question.

_____ 1. The ductus deferens is also called the <u>vas deferens</u>.

_____

_____

_____ 2. Reproduction in simple life forms that requires no partner is described as <u>sexual</u>.

_____

_____

_____ 3. The bulbourethral glands are also called <u>Bartholin's glands</u>.

_____

_____

_____ 4. The urethra passes through the <u>corpus spongiosum</u> of the penis.

_____

_____

_____ 5. The region of the brain that controls the pituitary is the <u>hypothalamus</u>.

_____

_____

_____ 6. <u>Estrogen</u> is the first ovarian hormone produced in the menstrual cycle.

_____

_____

VII. Completion Exercise

Write the word or phrase that correctly completes each sentence.

1. The process of cell division that reduces the chromosome number by half is _____

2. The channel through which the testes descend into the scrotum is the _____

3. The individual spermatozoon is very motile. It is able to move toward the ovum by the action of its _____

4. The main male sex hormone is _____

5. In the male, the tube that carries urine away from the bladder also carries sperm cells. This tube is the _____

6. The hormone that causes ovulation in females is the same hormone that stimulates cells in the testes to produce testosterone. In females, this hormone is called luteinizing hormone (LH); in males, it is called _____

7. The use of artificial methods to prevent fertilization or implantation of the fertilized ovum is called _____

VIII. Practical Applications

Study each discussion. Then write the appropriate word or phrase in the space provided.

Group A

The following patients were among those seen by a physician in a clinic that specialized in women's diseases.

1. Mrs. K, age 43, thought that the bleeding she was now experiencing might be associated with early menopause. The physician examined her and found a firm mass in the upper part of the uterus. This small rounded part located above the level of the tubal entrances is the _____

2. Ms. C, age 26, was seen for a routine examination. A slide was made of the cells in the neck of the uterus to be examined for cancer. The scientific name for this lowest region of the uterus is _____

3. Mrs. J, age 47, came to the clinic with complaints of hot flashes and irregular bleeding. Her history included infrequent ovulation, as evidenced by her irregular menstrual cycles. The physician did a biopsy of the uterine lining as an aid in determining whether the bleeding was due to cancer. The lining of the uterus is the _____

4. Mrs. K came in for evaluation of apparent sterility. After 12 years of marriage, she had no children. Her examination included an examination of the oviducts. Another name for these structures is uterine tubes or _____

Group B

The following patients were seen by a physician in a urology clinic for men.

1. Mr. T, age 32, complained of the presence of a lump on one testis. He had discovered it during his regular self-examination, which he did while showering. His physician ordered further tests to determine whether the lump was malignant. The testes are contained in the external genital structure called the _____

2. Mr. C, age 37, requested a surgical procedure that would render him sterile. This procedure, in which a segment of the ductus (vas) deferens is removed, is called a(n) _____

3. Mr. J, age 58, came in to his physician complaining of difficulty in urinating. Examination revealed an enlargement of a gland that is often affected in this way in men past middle age. This is the _____

4. Mr. J had a temperature of 39°C (102°F). In addition to the enlargement of the gland surrounding the first part of the urethra, there was also infection that involved the tortuous muscular tubes that contribute to the semen. These are the _____

IX. Short Essays

1. Name the pituitary hormones that are active in reproduction and describe their actions in both males and females.

   _____

   _____

   _____

   _____

2. Describe the function of semen and name the glands that contribute secretions to semen.

   _____

   _____

   _____

   _____

3. Define contraception and describe how several common methods of contraception work.

   _____

   _____

   _____

   _____

Memmler, RL, Cohen, BJ, Wood, DL. *STUDY GUIDE FOR STRUCTURE AND FUNCTION OF THE HUMAN BODY*, 6/e, © 1996, Lippincott-Raven Publishers

Development and Heredity

I. Overview

Pregnancy begins with fertilization of an ovum by a spermatozoon to form a *zygote*. Over the next 38 weeks, the offspring develops first as an *embryo* and then as a *fetus*. During this period, it is nourished and maintained by the *placenta*, formed from tissues of both the mother and the embryo.

Childbirth or *parturition* occurs in four stages, beginning with contractions of the uterus and dilation of the cervix. Subsequent stages include delivery of the infant, delivery of the afterbirth, and control of bleeding.

Milk production, or *lactation*, is stimulated by the hormones prolactin and oxytocin from the pituitary gland. Removal of milk from the breasts is the stimulus for continued production.

Heredity is determined by independent units called *genes* that are contained in the *chromosomes* of each cell. The chromosomes are passed from parents to offspring through the *germ cells* formed by the process of *meiosis*. Genes direct the formation of *enzymes*, which in turn make possible the chemical reactions of metabolism. Genes may be classified as *dominant* or *recessive*. If one parent contributes a dominant gene, then any offspring who receives that gene will show the trait. Traits carried by recessive genes may remain hidden for generations and be revealed only if they are contributed by both parents. Some human traits are determined by a single pair of genes (one gene from each parent), but most are controlled by multiple pairs of genes acting together.

II. Topics for Review

A. Pregnancy
  1. Fertilization
  2. Embryo
  3. Placenta
  4. Amniotic sac
  5. Fetus
B. Childbirth
  1. Stages
  2. Multiple births
C. Lactation

D. Heredity
  1. Chromosomes
    a. Distribution of chromosomes to offspring
  2. Genes and their functions
    a. Dominant and recessive genes
    b. Sex determination
    c. Sex-linked traits
    d. Multifactorial inheritance
    e. Factors influencing gene expression
  3. Mutation

III. Matching Exercises

Matching only within each group, write the answers in the spaces provided.

Group A

perineum   gestation   progesterone
amniotic fluid   implantation   placenta
embryo

1. The hormone produced by the corpus luteum that prepares the endometrium for the fertilized ovum
   _____

2. The flat, circular structure that serves as the organ for nutrition, respiration, and excretion for the fetus
   _____

3. Term for the developing organism from fertilization of an ovum until the third month of growth
   _____

4. The period of development in the uterus
   _____

5. The substance that surrounds the developing offspring and serves as a protective cushion
   _____

6. The pelvic floor in both males and females
   _____

7. Attachment of the fertilized egg to the lining of the uterus
   _____

Group B

lactation   dilation   prolactin
chorionic gonadotropin parturition   afterbirth
umbilicus   fetus

1. The process of giving birth to a child; labor
   _____

2. A hormone produced by embryonic cells that maintains the corpus luteum early in pregnancy
   _____

3. Widening of the opening of the cervix during labor _____

4. The scientific name for the navel _____

5. The secretion of milk _____

6. Term for the developing child from the third month until birth _____

7. The placenta, together with the membranes of the amniotic sac and most of the umbilical cord, as they are expelled after delivery of a child _____

8. The pituitary hormone that stimulates the secretion of milk by the mammary glands _____

Group C

colostrum   zygote   embryology
vernix caseosa   oxytocin   umbilical cord
abortion

1. The first mammary gland secretion to appear during lactation _____

2. The cheesy material that protects the skin of the fetus _____

3. Loss of a fetus before the 20th week of gestation _____

4. The study of early development _____

5. The structure that supplies nutrients and oxygen to the fetus and carries away waste _____

6. A fertilized egg _____

7. The pituitary hormone that promotes the ejection of milk from the mammary glands _____

Group D

genes   meiosis   carrier
chromosomes   dominant   mutation
multifactorial

1. Independent units of heredity _____

2. Term that describes a gene that is always expressed if present _____

3. The threadlike bodies in the nucleus that contain the genes _____

4. A change in the genetic material of a cell _____

5. Term for a person who has a recessive gene that is not expressed _____

6. The type of cell division that forms the gametes _____

7. The form of inheritance in which traits are determined by two or more pairs of genes acting together

_____

Group E

DNA	enzyme	heredity
XX	sex-linked	XY
mutagenic		

1. The complex chemical that makes up the chromosomes

_____

2. Term that describes traits transmitted by genes

_____

3. A protein that promotes a chemical reaction within a cell

_____

4. The sex chromosomes that appear in male cells

_____

5. Term for a trait carried on the X chromosome, such as hemophilia

_____

6. The sex chromosomes that appear in female cells

_____

7. Term for an agent that is known to produce changes in the genetic material of a cell

_____

IV. Multiple Choice

Select the best answer and write the letter of your choice in the blank.

1. A surgical cut and repair of the perineum to prevent tearing of tissue during childbirth is a(n)

   a. hysterectomy
   b. mastectomy
   c. episiotomy
   d. eclampsia
   e. hernia

1. _____

2. Twins that result from two different ova being fertilized by two different sperm cells are described as

   a. monozygotic
   b. identical
   c. fraternal
   d. preterm
   e. umbilical

2. _____

3. The second stage of labor includes

   a. expulsion of the afterbirth
   b. the onset of contractions
   c. implantation of the embryo
   d. passage of the fetus through the vagina
   e. expulsion of the placenta

3. _____

4. The term *viable* is used to describe a fetus that is   4. _____

    a. spontaneously aborted
    b. delivered by cesarean section
    c. developing outside the uterus
    d. capable of living outside the uterus
    e. stillborn

5. The person who is credited with the first scientific study of heredity was an Austrian monk named   5. _____

    a. John Down
    b. Gregor Mendel
    c. Bernard Sachs
    d. Alexander Wilson
    e. H. F. Klinefelter

6. Genes act by controlling the manufacture of   6. _____

    a. enzymes
    b. sugars
    c. DNA
    d. mutagens
    e. oxygen

7. In meiosis the chromosome number of a human cell is changed from   7. _____

    a. 23 to 46
    b. 36 to 18
    c. 100 to 50
    d. 46 to 92
    e. 46 to 23

V. Labeling

For each of the following illustrations, write the name or names of each labeled part on the numbered lines.

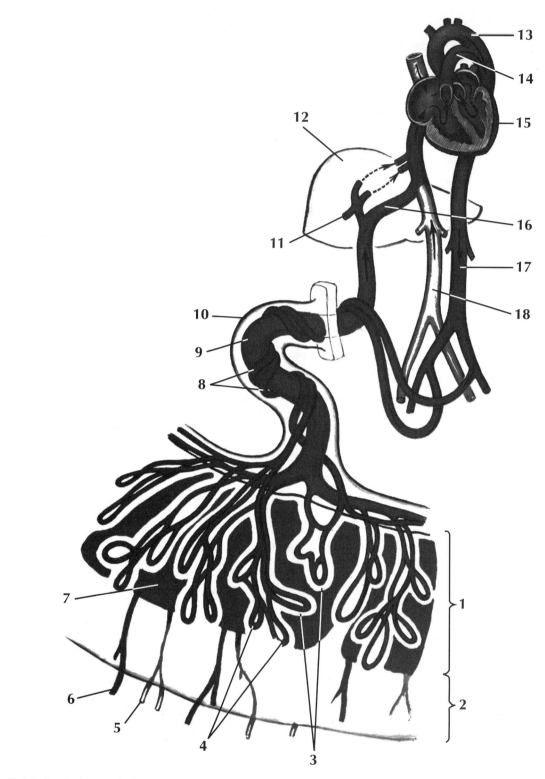

Fetal circulation and placenta

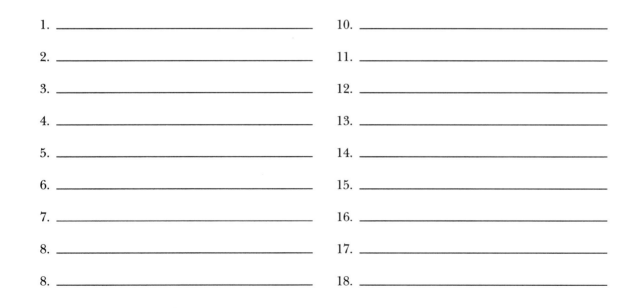

1. ———————————

2. ———————————

3. ———————————

4. ———————————

5. ———————————

6. ———————————

7. ———————————

8. ———————————

8. ———————————

10. ———————————

11. ———————————

12. ———————————

13. ———————————

14. ———————————

15. ———————————

16. ———————————

17. ———————————

18. ———————————

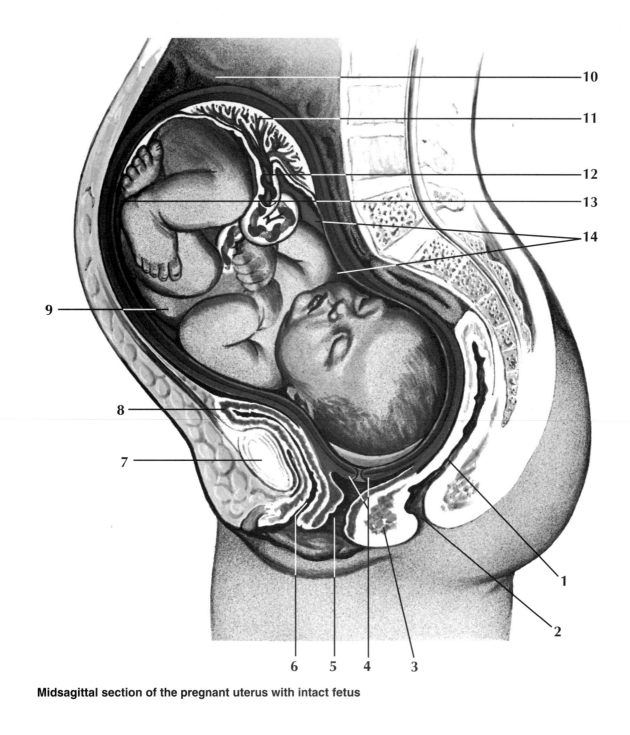

Midsagittal section of the pregnant uterus with intact fetus

1. _____   4. _____

2. _____   5. _____

3. _____   6. _____

7. _____

8. _____

9. _____

10. _____

11. _____

12. _____

13. _____

14. _____

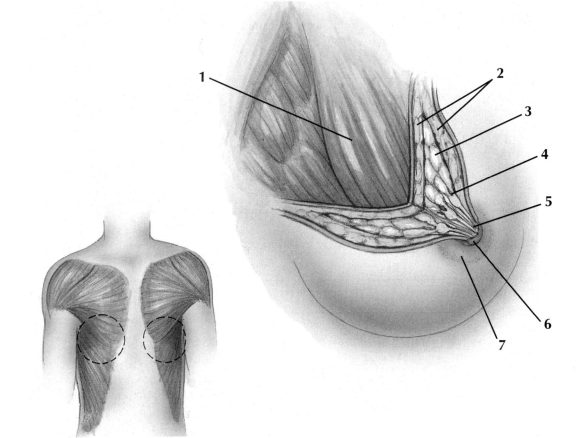

Section of the breast

1. _____

2. _____

3. _____

4. _____

5. _____

6. _____

7. _____

VI. True-False

For each question, write T for true and F for false in the blank to the left of each number. If a statement is false, correct it by replacing the underlined term and write the correct statement in the blanks below the question.

_____ 1. For the first 8 weeks of life in the uterus, the developing child is referred to as a <u>fetus</u>.

_____

_____

_____ 2. The beginnings of all body systems are established during the <u>first trimester</u>.

_____

_____

_____ 3. The afterbirth is expelled during the <u>third stage</u> of labor.

_____

_____

_____ 4. <u>Fraternal twins</u> develop from a single fertilized egg.

_____

_____

_____ 5. There are <u>two</u> umbilical arteries.

_____

_____

_____ 6. Blood is carried from the placenta to the fetus in the <u>umbilical arteries</u>.

_____

_____

_____ 7. The <u>Y chromosome</u> is the larger of the two sex chromosomes.

_____

_____

_____ 8. The sex of the offspring is determined by the sex chromosome carried in the <u>sperm</u>.

_____

_____

_____ 9. Every cell in the body except the germ cells contains <u>46</u> chromosomes.

_____

_____

_____ 10. A gene that must be inherited from both parents to appear is described as <u>dominant</u>.

_____

_____

VII. Completion Exercise

Write the word or phrase that correctly completes each sentence.

1. By the end of the first month of embryonic life, the beginnings of the extremities may be seen. These are four small swellings called _____

2. The cell formed by the union of a male sex cell and a female sex cell is called a(n) _____

3. The science that deals with the development of the embryo is called _____

4. The "bag of waters" is a popular name for the membranous sac that encloses the fetus. The scientific name for this structure is _____

5. The medical term for the loss of an embryo or fetus before the 20th week of pregnancy is _____

6. The mammary glands of the female provide nourishment for the newborn through the secretion of milk; this is a process called _____

7. An infant born before the organ systems are mature is considered immature or _____

8. The cutting of the perineum to reduce tearing during childbirth is called a(n) _____

9. A change in a gene or chromosome is called a(n) _____

10. The number of chromosomes in each human germ cell is _____

VIII. Practical Applications

Study each discussion. Then write the appropriate word or phrase in the space provided.

Group A

1. Mrs. G complained of soreness and discomfort of the breasts following the birth of her baby. The physician diagnosed her disorder as inflammation of the breasts, which are also called _____

2. Because of episodes of hemorrhage during her pregnancy, Mrs. M was hospitalized. There was a possibility that in Mrs. M's case the tissue that nourishes the developing fetus was attached to the lower part of the uterus instead of the upper part. This specialized tissue of pregnancy is the _____

3. Denise had missed several periods and suspected she was pregnant, but she did not consult a physician until her pregnancy was in its second trimester. The gestation period is 38 weeks, but if pregnancy is dated from the last menstrual period it is measured as _____

Group B

These patients were seen in a pediatric clinic.

1. A black child, about 4 years of age, was brought to the hospital with a history of swelling and pain in the joints of his hands and feet. Blood studies showed crescent-shaped red blood cells typical of a hereditary disease called sickle cell anemia. This disease is caused by a gene that must be received from both parents in order to appear. Such a gene is described as _____

2. Baby D's face was round; her eyes were close-set and slanted upward at the sides, an appearance typical of a genetic disease called Down syndrome. This condition is usually caused by an extra strand of DNA in the cells. These strands, which carry the genes, are called _____

3. Baby D's condition was probably caused by an unexplained change in the genetic material of her cells during development. Such a spontaneous change is called a(n) _____

4. Mrs. E began to suspect by the time he was 2 years old that her son had red-green color-blindness. This trait is carried on the X chromosome and is passed from mothers to sons. Any gene that is carried on a sex chromosome is described as _____

5. Mr. and Mrs. S noticed that even though they both had light brown hair their five children had hair color ranging from dark blond to brown. Hair color, as well as many other human traits, is determined by several gene pairs acting together. Traits determined by more than one gene pair are described as _____

6. Mr. and Mrs. C were delighted to give birth to their first child, a girl. The sex of offspring is determined by a pair of chromosomes named X and Y. The genetic makeup of all females is _____

7. Mrs. F and her husband received genetic counseling. They had one normal child and one child that lacked normal skin pigmentation, an albino. They wanted to know whether there was a possibility that a third child would be an albino also. Neither parent was an albino but each had apparently passed on a gene for the trait to one of their children. A person who has a gene that does not appear but can be passed on to offspring is called a(n) _____

IX. Short Essays

1. Name the pituitary hormones that are involved in lactation and describe what they do.

_____

_____

_____

_____

2. Explain how a child can be born with a specific trait that neither parent has.

_____

_____

_____

_____

3. Some traits in a population show a range instead of two clearly alternate forms. List some of these traits and explain what causes this variety.

_____

_____

_____

_____

Memmler, RL, Cohen, BJ, Wood, DL. *STUDY GUIDE FOR STRUCTURE AND FUNCTION OF THE HUMAN BODY*, 6/e, © 1996, Lippincott-Raven Publishers

Biologic Terminology

CHAPTER 22

I. Overview

Biologic terminology is the special language used worldwide by persons in scientific occupations. Many terms used today originated from Latin or Greek words, but some have come from more recent languages such as French and German. New words are being added constantly as discoveries are made and the need for words to describe them arises. Because scientific knowledge grows in different places at the same time, there may be two or even more terms in use that mean the same thing. Efforts are always being made, however, to standardize the terminology so that people all over the world will "speak the same language."

Not only does biologic terminology have universal application, but there are other advantages to its use. Often someone will say, "Why not use simple plain English?" The fact is that often there is no common word that is as precise as the scientific term. Moreover, one word or perhaps two can do the work of several sentences in descriptive force and accuracy. Biologic terminology is a kind of shorthand; scientists should be so familiar with it that it becomes a "second language" with which they feel completely at ease.

Most biologic words are made up of two or more parts. The main part, to which the other parts are attached, is called the root. These other parts include prefixes, which come before the root, and suffixes, which follow the root. A combining form is a root with a vowel added to make pronunciation easier. If more than one root or combining form makes up the word, it is a compound word. Take the time to divide each biologic word into its parts and then look up the meaning of each part, studying each as you go; you will soon add many words to your vocabulary. Then if you practice saying the word, you will feel at ease with this terminology. Here are some examples:

1. *Hypothermia* (hi-po-THER-me-ah): below-normal body temperature, usually due to excessive exposure to cold weather or icy water
   a. prefix: *hypo* = below normal
   b. root: *therm* = heat
   c. suffix: *ia* = condition or state of being
2. *Cardiopulmonary* (kar-de-o-PUL-mo-nar-e): related to heart and lungs
   a. combining form: *cardio* = heart
   b. root: *pulmon* = related to the lungs
   c. suffix: *ary* = pertaining to
3. *endometrial* (en-do-ME-tre-al): pertaining to the lining of the uterus
   a. prefix: *endo* = within
   b. root: *metr* = uterus
   c. suffix: *ial* = pertaining to
4. *megakaryocyte* (meg-ah-KAR-e-o-site): a giant bone marrow cell that releases blood platelets
   a. prefix *mega* = large
   b. combining form *karyo* = nucleus
   c. root *cyte* = cell

II. Topics for Review

1. common word roots and combining forms, such as
   abdomin-, abdomino-
   aden-, adeno-
   arthr-, arthro-
   bio-
   bronchi-, broncho-
   cardi-, cardio-
   cephal-, cephalo-
   chole-
   chondr-, chondro-
   cleid-, cleido-
   cost-
   cyt-, cyto-
   derm-, derma-
   enter-, entero-
   gastr-, gastro-
   gyn-, gyne-, gyneco-
   hem-, hema-, hemato-, hemo-
   hist-, histo-, histio-
   hyster-, hystero-
   idio-
   lact-, lacto-
   leuc-, leuk-, leuko-
   nephr-, nephro-
   neur-, neuro-
   psych-, psycho-
   somat-, somato-
   vas-, vaso-
2. common prefixes (beginnings of words), such as
   a-, an-
   ab-

circum-

contra-

di-

ex-

infra-

inter-

intra-

macro-

meg-, mega-, megalo-

met-, meta-

micro-

neo-

poly-

post-

semi-

sub-

trans-

tri-

uni-

3. common suffixes (word endings), such as

-ase

-blast

-cele

-ectasis

-esthesia

-ferent

-gen

-geny

-gram

-graph

-logy, -ology

-meter

-oid

-phagia, -phagy

-phil, -philic

-pnea

-scope

-tropic

4. common adjective endings, such as *-ous* and *-al*

5. common noun endings including *-us* and *-um*

III. Matching Exercises

Matching only within each group, write the answers in the spaces provided.

Group A

prefix	suffix	root
-ous	-logy	a-
compound word	combining form	

1. The foundation of a word   _____

2. A word that contains two or more word roots or combining forms   _____

3. The part of a word that precedes its root and changes its meaning   _____

4. A word ending used to change the meaning of the word root   _____

5. An ending that indicates the adjective form   _____

6. A word root followed by a vowel to make pronunciation easier   _____

7. A suffix that means *study of*   _____

8. A prefix that denotes absence or deficiency   _____

Group B

psych-	abdomin-	trans-
cyt-	hema-	somat-
hist-	aden-	neo-

1. A root that indicates the belly area   _____

2. A word root that means *gland*   _____

3. A root that shows relation to a cell   _____

4. A word root for tissue   _____

5. A root that shows relation to the mind   _____

6. A word part that means *blood*   _____

7. A word root that indicates the body   _____

8. A prefix that means *new*   _____

9. A prefix that means *through* or *across*   _____

Group C

arthr-	dorso-	dent-
-esthesia	infra-	-blast
meg-	heter-	viscer-

1. A prefix that indicates excessively large   _____

2. A suffix that means an *immature cell* or *early stage*   _____

3. A root that means *tooth*   _____

4. A root that shows relation to a joint _____

5. A root that refers to internal organs _____

6. A combining form that refers to the back _____

7. A suffix that refers to sensation _____

8. A prefix that means *other* or *different* _____

9. A prefix that indicates that a part is located below _____

Group D

-us   -al   encephal-
leuko-   -genic   -ase
cardi-   erythr-   ab-
-ism

1. A suffix that means *producing* _____

2. A suffix for an enzyme _____

3. A root that means *heart* _____

4. A root that means *brain* _____

5. A word part that means *red* _____

6. A word part that means *white* _____

7. A prefix that means *away from* _____

8. An ending that shows the adjective form _____

9. An ending for the noun form of a word _____

10. An ending that means *state of* _____

Group E

Combine word parts from the list below to form words that match each of the
following definitions. Write the correct words in the blanks.

hemo-, hemat-, or hemato-   -scope   -logy
-costal   -um   oste- or osteo-
inter-   -lysis   micro-
broncho-   bio-   -cellular
chondri-, or chondro-   peri-   intra-
cyto-, or -cyte

1. The study of living things is called _____

2. The connective tissue membrane covering a bone is the _____

3. An instrument for studying objects too small to be seen with the eye alone is a(n) _____

4. The scientific study of cells is known as _____

5. An instrument for studying the air passageways of the respiratory system is a(n) _____

6. The space between the ribs is described as _____

7. A cartilage cell is a(n) _____

8. The study of blood and its constituents is _____

9. The study of organisms too small to be seen with the eye alone is _____

10. The word that means between cells is _____

11. The word that means inside of or within a cell is _____

12. The connective tissue membrane that covers cartilage is _____

13. The dissolution or disintegration of blood cells (especially red blood cells) is called _____

14. Destruction or dissolution of body cells may be called _____

15. A mature bone cell is a(n) _____

IV. Completion Exercise

Write the word or phrase that correctly completes each sentence.

1. A prefix that indicates very small size is _____

2. Words that refer to an instrument for recording end with _____

3. The visible record produced by a recording instrument is indicated by a word ending in _____

4. A prefix that denotes *below* or *under* is _____

5. To show that something is outside or is sent outside, use the prefix _____

6. To indicate that there are three parts to an organ, begin the word with the prefix _____

7. A prefix that means *across, through,* or *beyond* is _____

8. To indicate a vessel use the root angi-, or _____

9. The prefix that shows something is within the structure is _____

10. The noun form of the adjective *mucous* is _____

11. Suffixes that indicate the process of eating or swallowing include _____

12. A root that means *joint* is _____

13. The prefix that means *away from* is _____

14. A root that means *skull* is _____

15. A root that means *muscle* is _____

16. A two-letter prefix that means *absence* or *lack* of is _____

17. A suffix that means *dilation* or *expansion of a part* is _____

18. An agent that produces or originates is indicated by the suffix _____

19. The word root for tissue is _____

20. Prefixes that mean *excessively large* include _____

V. Practical Applications

Study each discussion. Then write the appropriate word or phrase in the space provided.

1. Baby John was brought to the clinic by his observant mother because one of his eyes did not seem normal. The doctor noted that there was unilateral enlargement of the right pupil and that this uniocular condition would require laboratory investigation. The prefix *uni-* means _____

2. Mrs. B was admitted for treatment of an injured hand. The admitting intern noted that examination of the metacarpal bones showed possible fractures. The prefix *meta-* means _____

3. Ms. C was examined in the outpatient department. It was noted that she had circumoral pallor and that this pallor was circumscribed. The prefix *circum-* means _____

4. Baby M was born with an enlarged head due to accumulation of fluid within the skull. This condition is termed hydrocephalus. The word part *hydro-* in this name means _____

5. Mrs. A was admitted for surgery because of bleeding from the uterus. One word root for uterus is *metr-*; another is _____

6. A first-aid measure everyone should be able to perform involves the heart and the lungs. This type of resuscitation is described by the compound word for heart and lungs, which is _____